THE NATURAL HISTORY OF
BRITAIN'S COASTS

ERIC SOOTHILL AND MICHAEL J. THOMAS

THE NATURAL HISTORY OF BRITAIN'S COASTS

ERIC SOOTHILL AND MICHAEL J. THOMAS

GUILD PUBLISHING LONDON

This edition published 1988 by
Book Club Associates
by arrangement with
Blandford Press

Typeset by Asco Trade Typesetting Ltd, Hong Kong
Printed in Hong Kong
by South China Printing Co.

Contents

Acknowledgements

The authors would like to thank the following organisations and individuals for help in the preparation of this book:

Nature Conservancy Council (Mike Gash, Bob Davis, Malcolm Rush): Wildfowl Trust; RSPB; County Naturalists' Trusts, in particular Devon, Dorset, Kent, Durham, West Wales, Gwent, Glamorgan, and Manx Conservation Trust; Bobby Tulloch, Brian Gadsby and Myrfyn Owen for photographic contributions; David Stark, Dr Sarah Wanless, Peter Kincar and Ian Cumming for supplying valuable statistics; Ann Thomas, for patient and extensive research throughout the preparation period.

Eric Soothill would like to extend his personal thanks to the following people, whose homely accommodation and friendliness made his journey around the Scottish coast, during the summer of 1984, the more pleasurable:

Ella MacGeachy, Anstruther; Mrs M.L. Cowie, Helmsdale; Mrs E. Challinor, Tongue; Mrs C. Nicholson, Dunnet; Edith Fraser, Dunbeath; Lillian Brown, Golspie; Miss R. Capaldi, North Berwick; the Proprietor, Sunnybrae Hotel, Nairn; the staff, Marine Hotel, Muchalls.

Picture Credits

All photographs are by Mike Thomas and Eric Soothill, except for:
Brian Gadsby, Wildfowl Trust (p. 167);
Wildfowl Trust, Slimbridge (p. 30);
M. Owen, Wildfowl Trust (p. 207);
Bobby Tulloch, Shetland (pp. 15, 16, 17, 31, 72 and 73);
Richard Soothill (p. 108).

1

SEA-CLIFFS,
ROCKY ISLANDS AND
ROCKY SHORES

Introduction

Cliffs and rocky islands dominate the coastal landscape of Britain, providing some spectacular natural features and a wide variety of rock types. Britain is internationally famous for its large seabird colonies, some of which are the largest in the world. The most notable sites are situated along the west coast of England and Wales, together with almost all of Scotland. Along with providing nesting ledges for the seabirds, cliffs and rocky shores have their own distinct flora, which at certain times of the year can make an area of coast breathtakingly beautiful.

The majority of our sea-cliffs are composed of hard rocks like granite and sandstone which are able to resist the continuous impact of waves. Granite is the characteristic rock of the Cornish coast, the Scilly Isles and a few of the smaller western isles of Scotland like Ailsa Craig and Rhum. Sandstone, on the other hand, often produces spectacular effects in the form of arches, stacks and gullies. It is common along the Pembrokeshire coast, including the offshore island of Skokholm. The extensive Old Red Sandstone of Caithness and Orkney weathers to create steep cliffs full of ledges suitable for nesting seabirds.

Softer rocks like limestone and chalk are prone to continuous erosion, often resulting in the formation of high vertical structures. However, these are occasionally far too unstable to support a rich wildlife. Fine examples occur along the Durham coast and at several locations in Wales, such as the Ormes at Llandudno and parts of the Tenby Peninsula in the south. Chalk is mainly confined to eastern and southern England, as far west as Dorset. These are also the regions where even more extensive erosion takes place, because much of the coastline is composed of soft sand and clay banks. Spectacular landslips occur from time to time and in several areas walking along the cliff-top is prohibited.

There can be little doubt that our coastline is a superb visual attraction and as a result is prone to considerable disturbance throughout most of the year. In addition, several holiday centres have sprung up within easy reach of cliffs and, since access to the beach is then expected, erosion is un-

Distinctive rock structure in Old Red Sandstone along the Pembrokeshire coast.

Duncansby Head – the north-east tip of mainland Britain. The pounding waves have carved castle-like stacks in the red sandstone.

Chalk cliffs in west Dorset.

avoidable. The resulting loss of plant variety and disturbance to nesting birds are unfortunate consequences of which most of the public remain unaware. Such a situation exists, for example, on the cliffs leading down to Durdle Door in Dorset.

Changes in agricultural practices in recent years have also become a threat to many coastal sites. In the past, the tradition of grazing cliff-top pastures allowed a typical sea-cliff flora to spread over significant areas inland. This was very much the case in areas like Cornwall, Devon, West Wales and much of Scotland, but there is now an increasing change to arable farming. Consequently, the widespread destruction of cliff-top plant communities is taking place and this in turn affects the animal population of the habitat as well.

Reduced grazing by sheep and rabbits allows a coarser vegetation to develop in the form of dense scrub, and this makes an area less attractive to some butterfly species. It is said that the now extinct Large Blue (*Maculinea arion*) was a victim of such habitat-loss.

Different factors place our seabird colonies under threat. The most serious of these by far is the possibility of oil spillages from tankers and oil-wells offshore. Some areas like the east coast of England and Scotland are more susceptible than others because of the oil-field concentrations in the North

North Cornish coastline near Tintagel.

Rocky coastline of Dyfed near Moylegrove, west of Cardigan.

Sea. Wherever oil is transported by sea in bulk, however, the potential threat remains. In recent years, the West Wales coastline, including the off-shore islands of Skomer, Skokholm and Grass-holm, has been subjected to spills from tankers approaching Milford Haven. The most recent spillage during the summer of 1985 was serious enough, with several thousand birds being killed or injured; but a major disaster would have a catastrophic effect on the huge bird colonies occupying the area.

Egg-collectors still pose a significant threat even though the laws protecting birds have been tight-ened in recent years. Peregrines remain the main target for thieves, and the young chicks too are vulnerable, with falconers willing to pay large sums of money for them. The careless introduction of predators like foxes and rats to new territories would devastate a breeding colony. The continued success of many offshore islands as breeding sites is without doubt dependent on keeping such predators away.

Fortunately, the remoter islands and indeed many rocky coastlines on the mainland are generally inaccessible to humans. Nevertheless, there remain several sensitive areas under threat from man's activities, and long-term effective management is essential in future years if the beauty of our wonderful coastline is to be maintained.

Flora and Fauna

One of the finest and most spectacular sights at the coast in spring and summer is a noisy colony of seabirds using up almost every available ledge on a cliff. Although visitors to Scotland are the most likely to encounter such a scene, parts of Wales, the south-west and northern England also have a great deal to offer.

On the more vertical and inacccessible cliffs, guil- lemots and razorbills are the most conspicuous; the former, especially, tend to occur in tightly-packed groups on ledges with little room to lay their single pear-shaped egg. In much of Scotland, the black guillemot is just as likely to be seen, although it does breed in small numbers in parts of Anglesey, Cumbria and the Isle of Man. Often sharing the same cliff-face will be kittiwakes, which have the ability to

Seabird colonies at the Ord of Caithness.

Section of a gannet colony at Hermaness, Shetland.

'cement' their nests of grass and seaweed to almost vertical rock. Fulmars prefer a more substantial and grassy ledge for nesting. Like the kittiwakes, they make use of plentiful up-draughts to glide effortlessly to and from their nests.

Some cliffs are not too steep and are covered with a thick soil layer on which various grasses develop. Occasionally rabbits inhabit such locations, often in association with puffins and possibly manx shearwaters on the remoter islands. Indeed, some of the largest colonies of nesting seabirds are located on small islands. Skomer in Pembrokeshire boasts a shearwater population in excess of 100,000 pairs, with some 6,000 pairs of storm petrels breeding on nearby Skokholm. Further out to sea, Grassholm is noted for its large gannetry, regularly exceeding 20,000 breeding pairs. The coastline of Scotland is dotted with small islands, and it is not surprising that population densities of several seabird species are the highest in Britain. The coastline of Shetland alone is some 1,600 km (1,000 miles) in length, much of which is composed of sea-cliffs. This compares with the cliffed coastline of England and Wales which extends to some 4,400 km (2,700 miles) in total. It must be said, though, that the Farne

Islands, just off the Northumbrian coast, take some beating as far as accessibility is concerned. Visitors have the chance literally to walk amongst colonies of puffins, shags, kittiwakes, guillemots and razorbills. The islands are within easy reach of millions of people living in the highly-populated towns and cities of northern England and southern Scotland.

There are several species of seabird which cannot really be described as exclusive cliff-nesters. The great black-backed gull, the lesser black-backed gull and herring gull are just as likely to be seen in other coastal habitats during spring and summer. One good reason for nesting on, or simply visiting, a sea-cliff or rocky island on which the smaller seabirds nest is the supply of food that becomes available in the form of eggs and chicks. Tunnel-nesters have continually to 'run the gauntlet' when they return to their burrows after fishing expeditions. Marauding gulls are well aware of the fact that a harassed puffin is more than likely to give up its catch before reaching its burrow. The other underground nesters – the manx shearwaters – have to endure a different, but equally distressing, experience when they return at night from their fishing grounds out at sea. The larger gulls will prey on them and thousands of shearwaters are savagely mauled each year around our coast.

Sea-cliffs also provide nesting sites for some of our rarer birds of prey, golden eagles and peregrine falcons especially. Other cliff-nesters include the

Puffins (*Fratercula arctica*) at Muckle Flugga, Hermaness, Shetland.

Ground littered with remains of sea urchins and small crabs where gulls have fed.

raven and, where cliff top scrub has developed over the years, smaller songbirds also breed. Although mainly confined to northern Scotland, rock doves nest in colonies on ledges in caves, or in cliff holes. Rock pipits usually choose wild, rocky coastlines to build their nests of dried grass, normally in the smallest rock crevice, preferably near a good supply of small molluscs on which they feed. Rocky coasts also display a good seaweed variety in different stages of decomposition. Insects attracted to rotting seaweed are also eaten by rock pipits.

Cliffs are excellent vantage points from which to observe feeding seabirds offshore. Auks, cormorants and shags are predominantly underwater feeders, but spend a lot of time above water swimming around their feeding area. This gives cliff-top observers an excellent opportunity to view these birds at close quarters, and in some locations the chance to watch basking seals which regularly use the base of rocky cliffs for this purpose. One cave

Otters (*Lutra lutra*) on the rocky shore of Shetland.

used regularly by nesting choughs on Bardsey Island in North Wales has been named Seal's Cave, since in order to get to the nest ornithologists have carefully to negotiate one or two difficult ledges directly above grey seals swimming below. Visitors to the Farne Islands are usually treated to a 'round the island' cruise on the return journey, and, apart from the sheer delight of seeing the bird colonies in close-up, common seals are present in numbers on the rocky outcrops.

If the fauna of sea-cliffs is spectacular at certain times of year, then without doubt so is their flora. Rock crevices and ledges in the spray-zone need only trap the minimum amount of soil to allow salt loving species like rock samphire and rock sea-lavender to develop. Above this zone and especially where the slopes are gentler, maritime grassland grows, accompanied by thrift, sea campion and spring squill.

Where the salt-spray reaches the cliff-top, as usually occurs on the more exposed cliffs of western and northern Britain, the vegetation may include species more typical of saltmarshes, such as sea milkwort and sea plantain. Should seabird colonies nest in the area, soil enrichment takes place, allowing such species as sea beet and common orache to develop.

On the less exposed cliffs, where the influence of salt-spray is reduced, fewer maritime species occur. Such situations exist on the east coast of England, and since the cliffs in that region are mainly composed of sand and clay a greater diversity of flowering plants can be found.

Grazing by rabbits and possibly sheep also plays an important role in allowing several rare non-maritime species to flourish. This is certainly true of calcareous grasslands, like the Little and Great Ormes of North Wales. The latter even supports a herd of feral goats, which also helps with controlling the development of scrub and the consequential ruination of a unique habitat.

Below sea-cliffs, many micro-habitats exist, offering varying habitats to the rocky shore fauna. Acorn barnacles attach themselves to the lower rocks, well within reach of the tides. Predators like the dog whelk blend in with the barnacle-covered rock because of their dull white colouration. Empty barnacle shells often house minute bivalves and, just above the barnacle belt, in lichen-covered rock crevices, small, smooth-shelled and rough periwinkles spend their lives attached to the same rocks as beadlet anemones. Other anemones, like the snake-lock and daisy anemone, occur lower down the shore in shallow open pools. Many of these are preyed upon by the grey sea-slug which feeds solely on anemones.

Crabs are common on rocky shores, with the shore crab abundant in some areas, but hidden below large stones at low tide. There, too, will be found the common blenny and possibly a gunnel and bullhead, especially nearer the low-water mark. Common starfish also conceal themselves beneath rocks, together with their relatives the sea cucumbers and perhaps the occasional sea urchin.

Many of the univalve molluscs are far more conspicuous, with the edible periwinkles openly displaying themselves on bare rocks in between tides. However, the threat of dehydration prompts most to remain submerged or at least in the shelter of moist seaweed. Other periwinkle species also occur along different parts of the shore, as do the purple and grey topshells. On some shores, mussels are abundant, attaching themselves by strong threads to the rocks. Even then they fall prey to dog whelks and starfish which are both partial to a diet of fresh mussels.

Limpets, beadlet anemones, periwinkles, barnacles and sponges attached to the rock in a rock pool at low tide.

Mixture of seaweeds in a rock pool.

Barnacles, winkles, young mussels and sea lettuce on rocks at the foot of a cliff.

Empty topshells, winkle and dog-whelk shells are often inhabited by young hermit crabs – their larger parents preferring to live offshore in the larger shells of edible whelks.

The crustaceans of rock-pools include several species of prawns. The edible variety is mainly confined to the deeper pools of the lower shore.

Rock-pool vegetation is far more apparent on some shores than any animal species. Seaweeds extend throughout the various zones and play an important part in providing valuable cover for the animal life, none more so than the brown seaweeds which are by far the commonest. The wracks in particular can tolerate certain amounts of exposure. Of these, the furthest up the shore is invariably the channelled wrack which has the advantage of being able to trap moisture in the channels between its inrolled fronds and the rock surface. Along the middle-upper shore boundary, flat or spiral wrack occurs, and over most of the middle shore bladder and knotted wrack, both of which bear air-bladders for increased buoyancy. Towards the lower-middle shore, serrated wrack tends to form a mat over the rocks and is probably the commonest of the British seaweeds, extending way down to the lower shore into the deeper pools. Here exist the large tough seaweeds commonly known as the oarweeds, whose spreading holdfasts provide shelter for a range of small animals. Except during extreme low water spring tides, the oarweeds are rarely totally exposed and it is usually the bent stipes only that can be seen above the water, with the large fronds drooping into the water. Of the oarweeds, tangles have a substantial, smooth, cylindrical stipe culminating in long leathery fronds able to move with the changing currents. On the other hand, sea belt is wavy in appearance, has a shorter stipe and is unbranched.

Other tough lower-shore seaweeds include thong-weed, which has long strap-like branches developing from greenish, button-like growths attached to the rocks. Bootlace weed is somewhat similar in

Seaweed covering a rocky shore at low tide.

appearance, being olive-brown in colour, but it is unbranched, unlike thongweed.

Although less common than the brown species, the red seaweeds offer the greatest variety and are generally located from the middle shore downwards, since they are less tolerant to exposure. On some coastlines, the red seaweed *Gigartina stellata* is abundant and is similar in appearance to carra-

gheen, whose fronds however do not have in-rolled margins. Frequently occurring in the same situation is the pepper dulse, which shows a wide variation in colour from an olive-green in strong sunlight to dark reddish-purple in the deeper pools.

More common on the lower shore, usually attached to rocks or even the larger brown seaweeds, is dulse, which like carragheen is edible and was

Sea lettuce covering boulders below sea-cliffs at Penbryn Beach, Dyfed.

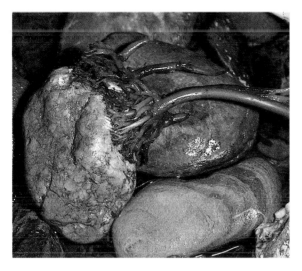

Holdfasts of sea belt (*Laminaria saccharina*) attached to rock pool stone.

Two species of lichen: bright orange *Xanthoria* and yellowish-grey sea ivory (*Ramalina siliquosa*).

once harvested in bulk around the coast of Britain, but more especially in Scotland. Purple laver continues to be a delicacy in South Wales and North Devon, and although classed as a red seaweed is brownish in colour.

The appearance of the shallower pools on rocky shores is enhanced by the pink colouration of the vegetation lining them. This is mainly caused by the encrusting red seaweed *Lithophyllum* and frequently the coral weed, which is much more tufted and ranges in colour from plum-coloured to a yellowish-pink or even white in bright sunlight.

Sea lettuce is a fairly common green seaweed on rocky shores, but is not always confined to rock pools. It often appears spread over rocks in the upper shore, especially near a freshwater outlet. *Cladophora* is a moss-like green seaweed of the middle and lower shores, often appearing in the same pools as some of the brown seaweeds, especially serrated wrack. *Enteromorpha* is another green seaweed which associates itself with other inhabitants of the rocky shore. Frequently found attached to coral weed, it also attaches itself to the slower-moving animals such as limpets.

Although there can be no question that seaweeds are the main plant colonisers of the rocky shore, there are also the lichens to consider. They are mainly found on rocks beyond the high water mark, in the spray zone. Perhaps the commonest, and certainly the most attractive, is the bright orange *Xanthoria*, whose beauty is in sharp contrast to the dull black of *Verrucaria*, regularly mistaken for black patches of tar. Where *Xanthoria* grows alongside the yellowish-grey tufts of sea ivory (*Remalina siliquosa*), the beauty of the rocky shore is enhanced, especially in late spring when thrift is in flower. The lichens can thus be regarded as a link between the seaweeds of the rock-pools and the flowering plants of the cliff-face, making the rocky shore one of the most fascinating habitats for the visitor.

Shag (*Phalacrocorax aristotelis*)
Length 76 cm (30 in)

Readily distinguished from the cormorant (*Phalacrocorax carbo*) by lack of white on cheeks and chin, it is also some 15 cm (6 in) smaller in stature, and when breeding displays a short erect crest. At close range the plumage is noticeably greenish-black (hence the alternative name Green Cormorant), whereas the cormorant is dark glossy black, and during the breeding season has a white thigh-patch.

Whilst the shag spends its entire life in a marine habitat, it must not be considered pelagic, as its search for food seems confined to inshore waters where it dives for fish. In open water sand-eels certainly seem to form the bulk of its diet; when in estuaries shore fish are taken in preference to flat fish; and in Scottish lochs herring are sometimes taken. Crustaceans are also eaten. The average duration per dive is about 50 seconds, with a maximum of just less than three minutes. In sheltered waters, rafts comprising several hundred birds may be seen. Occasionally shags are driven inland by autumn gales when they become stranded on lakes and reservoirs. Those not returning to their coastal habitat at the earliest opportunity will, in all probability, face a lingering death.

The shag usually breeds colonially, although scattered groups and even single nests are not uncommon. The nest is a large collection of material, mainly seaweed, but bracken, heather stems and various coastal plants are often incorporated; all of

Shags (*Phalacrocorax aristotelis*) on a rocky pyramidal island.

which is collected locally by the male bird. The nest is constructed on sea-cliff ledges, in caves, and on bare open rocky areas of isolated islands. Although occasionally clutches of eggs have been found at the end of February, it is true to say that egg-laying usually begins early in April, with most clutches being completed by early May. The three (or four) pale blue eggs are partially coated by a chalky deposit; incubation is undertaken by both sexes for a period of 30 days, the chicks being brooded for a further 14 to 16 days. Feeding is by regurgitation. At seven to eight weeks of age the young leave the nest, but will be fed by their parents for a further four or five weeks.

Apart from an assortment of croaks and an intermittent 'hissing', the shag is perhaps the quietest of all the European seabirds.

Distribution in Britain is chiefly along the north and north-west coasts of Scotland, and Scottish islands; coastal areas of Wales, Devon, Cornwall, and the Channel Islands; also Ireland in all but parts of the east coast. It breeds in Iceland; the Faeroe Islands; and from the Kola Peninsula south to Stavanger (Norway); Brittany; N.W. Iberia. Another race, *P.a. desmaresti*, occurs in the central Mediterranean; and along the coast of Morocco we find *P.a. riggenbachi*.

Herring gull (*Larus argentatus*).

Herring Gull (*Larus argentatus*)
Length 56 cm (22 in)

Much larger than the common gull (*Larus canus*) with which there is the possibility of confusion at distance. However, close inspection reveals a red spot at the base of the lower mandible of the yellow bill; a yellow iris; and pale flesh-coloured legs. It is the commonest of our coastal gulls and widely referred to by the layman as the 'seagull'. One problem of identification, common to all, is the differentiation between juvenile herring gulls and those of the lesser black-backed gull (*Larus fuscus*), an almost impossible task! Full adult plumage, of light grey back and wings, and the black-tipped primaries of an otherwise white plumage, is not fully attained until the fourth year.

There are few records of British-ringed herring gulls being recovered abroad. The birds from our major breeding colonies disperse to ancestral wintering areas along the coasts, where they regularly scavenge for food close to human habitation (rub-bish tips and even litter bins at coastal resorts); or forage on agricultural land; or search for items of food along the tide line, depending on local availability. During the breeding season egg-stealing and the killing of young birds (often their own species) is prevalent; or harassing members of the auk family (particularly puffins) until they release the catch of small fish from their bills.

Early in March the adults return to their colonial breeding grounds; the male birds having established territory guard it with great vigour from would-be intruders. Herring gull colonies are often large and always noisy. The nests are built by both sexes; when the site is a broad cliff-ledge or a sea-stack, the amount of material required is usually far more than would be needed for a ground nesting-site. Egg-laying begins in mid-April onwards, three being the normal clutch, and the eggs are quite indistinguishable from those of the lesser black-backed gull. Incubation continues for 28 to 33 days and usually commences with the laying of the second egg. The young fledge after about seven weeks, and during this time remain in or within the region of the nest.

The breeding distribution of the race *argentatus* extends from Iceland and the Faeroe Islands south to Britain and Ireland; north-western Europe and Scandinavia to the Baltic. The race *omissus* replaces *argentatus* in northern scandinavia and north-western Russia; hybridisation occurs quite freely where the two races overlap. In the north-eastern United States and throughout Canada we have *Larus smithsonianus*, the pink-legged North American species.

Great Black-backed Gull (*Larus marinus*)
Length 68.5 cm (27 in)

With a wingspan of about 1.5 m (5 ft), this must be Northern Europe's largest breeding gull. Plumage-wise the only possible confusion could be with the smaller (54-cm, 21-in) lesser black-backed gull (*Larus fuscus*), whose back and wings are slate-grey, not black. In both birds the rest of the plumage is white. The great has a slower wing-beat than the lesser. Bill very large, yellow with red spot towards tip of lower mandible. Legs flesh-coloured; this alone should be a sufficient feature to distinguish it from the lesser which has yellow legs in summer.

One of our most maritime gulls, the majority of which feed regularly on offal disposed of by fishing boats, it will occasionally scavenge at rubbish tips both coastal and inland. During the breeding season it is notorious for its vicious attacks on manx shearwaters (*Puffinus puffinus*), and puffins (*Fratercula arctica*), as they enter their breeding burrows. It also preys on many other species of birds and a variety of small rodents (including rabbits), and even sickly lambs; crustaceans, molluscs, insects and worms are also included in the diet.

The call is a deep guttural 'agh, agh, agh', and a short bass 'owk'. Each winter their numbers around our coasts are increased by visitors from abroad, particularly from northern Scandinavia and the Kola Peninsula (Russia).

In Britain, generally speaking, the great black-back tends to be a solitary nester, although large but scattered colonies do exist, for example North Rona

Great black-backed gull (*Larus marinus*) and chicks.

(Outer Hebrides), and Mullion Island (Cornwall). It is mainly a coastal breeder, although pairs do sometimes breed inland. When nesting it chooses such sites as the tops of stacks, small islands, and often in close proximity to colonies of other seabirds. The nest is merely a slight depression lined with dry grasses. Most clutches are complete by the end of April; the three pale buff to olive-brown eggs, which are blotched with varying shades of dark brown, are laid at intervals of two days. Incubation commences with the second egg and both birds share this duty for 26 to 28 days. The downy young leave the nest after two or three days, seeking cover in the nearby vegetation. Like other gulls they are fed by regurgitation. After seven or eight weeks the young have fledged and are now ready to leave the nesting grounds.

As a breeding species the great black-back is

widely distributed throughout Britain, with the coastal areas between Berwickshire in the east and south to Hampshire being the exceptions.

The most southerly breeding in Europe occurs in N.W. France and the Channel Islands. East of Britain they breed in coastal areas of Scandinavia, Finland, N. Estonia, and the Kola Peninsula (Russia). North and west of Britain they occur around the Icelandic coast, the Faeroe Islands, West Greenland, Bear Island, and the Atlantic coast of North America south to about the Delaware Bay, 40°N.

Kittiwake (*Rissa tridactyla*)
Length 39–41 cm (15–16 in)

Of similar size to the common gull (*Larus canus*). Adult birds have head, tails, and underparts white; back and wings grey. The lack of a white 'mirror' to its triangular black wing-tips prevents confusion with other gulls. The yellow bill, striking red gape, and short black legs are further aids to identity.

Outside the breeding season much of the time is spent wandering far out in the North Atlantic, where it feeds either by picking from the water surface whilst in flight, or diving from the surface if swimming. When fishing or following a ship in search of offal, the kittiwake usually plunge-dives whilst in flight. During the breeding season 90 per cent of the food taken is fish; sand-eels form a large part of the diet.

Early in spring the birds return to their traditional cliff nesting-sites, where even the narrowest of ledges is adequate; the collection of seaweed, grasses and other vegetation is firmly held together by a mixture of the bird's excreta and green algae. A small grass-lined cup is fashioned to take the eggs. Outside the breeding season kittiwakes are comparatively silent gulls, but the excitement generated as colonies gather to breed on our sea-cliffs gives rise to that loud and familiar onomatopoeic cry 'kitti-wa-a-k', 'kitti-wa-a-k', from which its common name is derived.

During early May two greyish-brown eggs are laid, which are blotched and spotted with darker shades of grey and brown. Both sexes share the incubation, which lasts for a period of 21 to 25 days. The young birds are fed in the nest for a further 40 days or so. An estimated population for Britain and Ireland would be approaching half a million birds (75 per cent of which occur around the coasts of

Kittiwakes (*Rissa tridactyla*) on nesting cliff.

Scotland), making the kittiwake our most numerous gull.

The kittiwake distribution in Northern Europe extends from Iceland to the northern coast of Russia, including islands in the Arctic Ocean, with colonies along the Norwegian coast; also the Faeroe Islands, Britain, Ireland, and France where a few small colonies occur on the coast of Brittany. On the other side of the North Atlantic there are colonies in Greenland, Baffin Island and southwards to the Gulf of St Lawrence and Newfoundland. The subspecies *R.t. pollicaris* occurs in the North Pacific in mixed colonies along with the red-legged kittiwake (*Rissa brevirostris*), but there seems not to be any interbreeding.

Fulmar (*Fulmarus glacialis*)
Length 47 cm (18½ in)

Although gull-like in appearance, the heavier rounded head and short thick neck, coupled with the external tubular nostrils, should be an adequate means of differentiation. Head, body, and somewhat fanned tail are whitish or ivory; contrasting with the light grey of back and upper wing surfaces (the latter being void of black tips). The name 'fulmar' is said to mean 'foul gull', no doubt from its habit of ejecting a nasty smelling oily liquid, which it often does if disturbed whilst incubating. Bill yellow with dark tubes. Legs and feet vary in colour from yellow to pale bluish-pink. The fulmar is quite unable to stand and rests its belly on the full length of its legs. When airborne it displays a masterly skill of gliding, either low over the water, or to and fro along the cliff-faces, only the occasional flap of its narrow wings being necessary.

Non-breeding birds seldom venture ashore, spending all their days wandering the oceans.

Although breeding does not commence until April, some fulmars are in occupation of their breeding-sites for ten months out of twelve, only vacating the steep slopes and ledges during September and October, and again for a period of two weeks or so prior to egg-laying. A single white egg is laid usually during May, a simple unlined scrape serving as a nest. Incubation is shared alternatively by both sexes for spells of four or five days, over a period of about 55 days. The young bird is fed by both parents. To obtain food (which in some cases seems to be very liquid) the chick places its bill inside that of the adult. The fledging period is about seven weeks and the young bird weighs more than its parents at that time. Fulmars usually colonise the broad ledges of cliff-faces; alternative sites being sand dunes, beaches, ruins, and even occupied dwellings. On one occasion we found a pair occupying a disused rabbit burrow in the middle of a tern colony, the unfortunate birds being severely mobbed with each venture to or from the nest.

In 1878 Shetland was the only breeding area in the British Isles; during the 100 years that followed the British population has risen to over 305,000 pairs, 88 per cent of which breed on St Kilda, the Outer Hebrides, the north-west coast of Scotland, Orkney, and Shetland.

The fulmar has been split into three races of which the large-billed *auduboni* breeds in Iceland, the Faeroes, northern Scotland and the Scottish islands, also locally in Norway, and sparingly in Brittany and Jersey. The small-billed *glacialis* breeds in Greenland, arctic Canada, and arctic Europe. The third race is *rodgersii* of the North Pacific.

Manx Shearwater (*Puffinus puffinus*)

Length 35 cm (13½ in)

A bird with sooty-black upperparts and white underparts. Wings long and narrow, tail short; and altogether of slender build. In British waters it cannot be mistaken for any other species of shearwater. During the daytime it can only be seen out at sea (except when forced inland by gale conditions); this makes close inspection of its black, slender, hook-tipped bill difficult. It has the tubular nostril common to its genus. The legs and webs of its dark feet are pinkish. Its calls are a selection of loud weird screams and gurgling sounds.

Manx shearwaters are colonial nesters, usually on small islands, and, although most seem to excavate burrows at not too great a distance from the sea, one exception is on the Isle of Rhum where breeding takes place 3 km (2 miles) inland, and in excess of 610 m (2,000 ft) above sea level.

Adult birds come ashore to breed in February, but the lengthy season extends to September or even October. Sometimes rock crevices can be utilised for nesting purposes but mostly the birds breed at the end of underground tunnels. These are from 0.5 to 2 m (1½ to 6 ft) in length, and sparsely lined to form a nest for the single white egg. Where the soil is not too tightly compressed the birds are well able to excavate their own tunnels. Prior to egg-laying, which commences in April, the shearwaters vacate the colonial site, and just like the fulmars spend about ten days out at sea. The only obvious reason for this sojourn is to get a high intake of food and so build up their reserves to see them through the 52 days' incubation period which is to follow. This is undertaken alternatively by both birds, often at irregular intervals (1–26 days), but averaging about six days. Visitations to the nest, either to incubate or later to feed the chick, are only carried out under cover of darkness. After seven to eight weeks of progressive feeding, the young bird attains a body-weight double that of an adult. At this stage it is deserted by its parents, and when about ten weeks old makes its first and often last journey to the sea. For now it must struggle on legs and feet that are not well designed for walking; in some cases a distance of many hundreds of metres must be undertaken before it reaches the sea and comparative safety. All this time it is under the constant threat of attack from large marauding gulls seeking an easy meal. Those that survive spend their first two years wandering out at sea before returning to home waters, but it is not until the fifth year that they are able to breed. The estimated breeding population of Britain and Ireland is in excess of 250,000 pairs. The bulk of these are accounted for in three main areas, Island of Rhum (Inner Hebrides); Skomer and Skokholm (off the Pembrokshire coast); also Blasket Isle and both the Great and Little Skellig (Co. Kerry).

Gannet (*Sula bassana*)
Length 90 cm (35 in)

With a combined body/bill length of 90 cm (35 in) and a wing span of almost 2 m (6 ft), the gannet is Britain's largest breeding seabird. Its familiar cigar-shaped body is easily identifiable, even at long range; adult birds have entirely white plumage shading to golden-buff on nape, head and crown; the wing-tips are black. Bill dagger-shaped and blue-grey with concealed nostrils; black at base, the black extending backwards above the pale yellow eye. The short legs, and feet, are greyish-black, and at close range it becomes apparent that male birds have turquoise green lines on their toes, whereas those on the female's are of a more yellowish-green. When in direct flight the strong and somewhat rapid wing-beats are often interrupted by short periods of gliding. In low flight out at sea it planes and swerves over the waves. It swims quite buoyantly with tail pointing upwards, but when on land it can only walk with a duck-like waddle, and seldom if ever ventures very far.

The gannet is a colonial nester, selecting cliff-ledges, grassy slopes above precipitous sea-cliffs, or small grass-covered islands. Birds begin to return to their breeding-sites as early as January, with numbers building up during the next two months. Nest-building is undertaken by both birds, and a large collection of dried vegetation, seaweed and feathers is accumulated, which is bonded together by excrement; some nests may be as high as 60 cm (24 in). Once a pair has mated they often occupy the same site for many years. A single pale blue egg is laid from April onwards; it soon becomes stained brown during the 44-day incubation period which is shared by both sexes. Gannets do not have brood patches; the egg is positioned beneath the bird's feet, being repositioned on top of the feet once it commences to 'chip'. Young gannets have a fledgling period of

about 90 days, and for the first 14 days they are brooded by their parents. The chicks are fed by regurgitation several times a day; the diet is one of fish, which is obtained by thrusting its bill down the adult bird's gape. To acquire fish the gannet plunge-dives, usually from a height of about 10 m (33 ft), but 30 m (98 ft) is not unusual. A speed of about 100 kph (62 mph) is reached during such a dive and the water is entered beak-first on gradually closing wings. Fish are caught at depths of up to 5 m (16 ft), grasped between the bill and brought to the surface before eating.

The total breeding population of the British Isles now stands at over 140,000 pairs with approximately 60,000 pairs inhabiting St Kilda (Outer Hebrides) making it the world's largest gannet colony. *S. bassana* is confined to the N. Atlantic, including Newfoundland, Gulf of St Lawrence, Iceland, Faeroe Islands, Norway, Ireland, Britain, Channel Isles and Brittany. The two other species, *S. capensis* and *S. serrator* of South Africa and Australia respectively, are considered by some authorities to be conspecific with *Sula bassana*.

Puffin (*Fratercula arctica*)
Length 30.5 cm (12 in)

The puffins, or sea parrots as they are sometimes called, make a return to our coastal waters during late February and March, having spent the winter months out at sea. From then until April, when breeding commences, they can be seen in rafts off-shore, making only the occasional visit to dry land. They are already paired and will soon spend more and more time ashore.

The courtship display is really a simple affair; both birds stand face to face and engage in their bill-shaking ritual interspersed with mutual head and neck nibbling, the male often making presentations of grass and feathers to his 'intended'.

Puffins are colonial breeders, choosing the grassy slopes of coastal cliffs, hillsides and islands as their habitat. The burrow, at the end of which in early May the female will lay her large single white egg on a sparse linging of grass, is excavated by both birds using their strong bills to good advantage as they loosen the soil or soft rock, before shovelling it away with their feet. Often the burrow of a rabbit or manx shearwater may be appropriated; in some colonies the nests are in cavities underneath rocks. During

Gannets (*Sula bassana*) breeding in close proximity.

the construction of burrows the birds become very vociferous, and it is amazing how many variations can be achieved from a voice which at other times is little more than a low growl.

Whilst it is generally accepted that both sexes share the incubation of some 40 to 43 days, it is the female who plays the major role by far. During this period few birds are to be seen on land in the vicinity of the burrows. On hatching, the chick is fed on small fish which the parents catch and hold crosswise in the beak, often up to 12 or even 18 at a time; the peak feeding periods are early morning and late afternoon. After about six weeks the chick, which is now very fat, is abandoned by its parents. During the next seven days or so without food the young puffin slims down and is now adequately feathered to leave its burrow and take to the sea; an event which takes place at night and usually occurs late in July or early August. It is now that the adult birds moult into a much drabber black and white plumage, and the bill becomes smaller and almost colourless as that colourful horny sheath is shed.

The birds also lose the horny red and grey triangle around each eye. Now adult birds and juveniles, with their smaller and darker beaks, can be seen in that typical low 'whirring' flight over the water.

The puffin breeds on both sides of the Atlantic from Novaya Zemlya to western Greenland. In the British Isles the main breeding areas occur along the west coast of Ireland, the coasts of northern Scotland including the Hebrides as well as parts of south Wales and south west England.

Common Guillemot (*Uria aalge*)
Length 42 cm (16½ in)

The head and back of this, the most slender of our auks, is dark chocolate-brown in the race *U.a.*

Puffins (*Fratercula arctica*).

albionis which breeds up to southern Scotland, but black in the race *U.a. aalga* which occurs in the rest of Scotland and northwards. The underparts are white in both races. The legs and webbed feet are dusky yellow, and the slender pointed bill black with a yellow gape. In the bridled variety a white line extends backwards from the white-ringed eye.

'Bridled' birds become more numerous in the northerly latitudes. For example they are uncommon in the south of England with fewer than 1 per cent, whilst some of the Shetland colonies have up to 26 per cent; in southern Iceland the percentage reaches 50–70.

With somewhere in the region of 580,000 breeding pairs in the British Isles (80 per cent of which are in Scotland) the guillemot outnumbers the razorbill (*Alca torda*) by about four to one. In Orkney an estimated 70,000 pairs inhabit North Hill on Westray; this is the largest British colony. By far the biggest colony in England is to be found on the cliffs at Bempton in Yorkshire.

Early morning visits to the breeding ledges are already occurring before Christmas, but at this time the birds seem very unsettled and vacate the ledges at the least intrusion; even a fishing vessel sailing close in will put them to flight. As May approaches

Guillemots (*Uria aalge*) on Ramna Stacks, Shetland.

parents; the pale green or blue egg is variously marked with scrawls and blobs of brown and soon becomes coated with the birds' excrement. This coating of guano, along with the egg's pear-shape and the increasing size of the air space at the wide end as incubation progresses, all contribute to make the egg stable on the bare rock ledge.

Young guillemots are fed a diet of small fish. After a period of three weeks or thereabouts they flutter, from the precarious breeding ledge, down to the comparative safety of the sea. Here, in company with their parents, who continue to feed them, they venture away from the foot of the cliff to deeper water. By August the breeding cliffs will once again be completely deserted.

Whilst the majority of guillemots are thought to winter close to shore there have been recoveries of ringed birds in Denmark, and Norway, these from colonies in northern Scotland.

World distribution: North Pacific from Oregon to Alaska and islands in the Bering Sea to Japan – the race *U.a. hyperborea*. Eastern North America to Labrador and West Greenland, also Iceland, Faeroe Islands, Norway and northern Scotland – *U.a. aalge* (nominate). Iberia to southern Scotland *U.a. albionis*.

Black Guillemot (*Cepphus grylle*)
Length 34 cm (13 in)

Unmistakable from our other breeding auks because of a broad white patch on wings; rest of plumage a uniform brown-black; legs and gape are bright red. A little larger than the puffin (*Fratercula arctica*), but somewhat smaller than both the razorbill (*Alca torda*) and the common guillemot (*Uria aalge*). In full winter plumage it retains the large white wing patch; but now the head and underparts are predominantly white; the sides of head, upperparts, rest of wing, and tail are mottled black and white. At this time of year the overall greyish-white appearance can cause confusion with other guillemots and also grebes.

The British Isles represent the black guillemot's most southerly breeding limits, so quite naturally one would expect it to be the least numerous of our auks. In fact the total breeding population is less than 10,000 pairs, with over 4,500 pairs breeding in Orkney and Shetland; the rest distributed mainly in the north and west of Scotland, Ireland, and the

the guillemots begin to pack every available space on the ledges, and there is constant commuting back and forth to the sea below. Nuptial display is soon in evidence, with mutual preening of head and neck between males and females, continual nervous bowing of the head with an accompanying harsh growl-like 'arrrrrr' rising to a crescendo!

By the end of May the single pear-shaped egg will have been laid on the bare rock ledge. Incubation lasts for four to five weeks and is shared by both

Isle of Man. Small numbers breed at the Ord of Caithness in eastern Scotland. The only English and Welsh sites are St Bees Head, Cumbria, and Anglesey respectively.

It is much more approachable than other auks, especially when resting on rock shelves which it does far more readily than either the razorbill or common guillemot. Also, unlike these two species, the black guillemot lays two eggs. The nest is a simple scrape lined with tiny pebbles and constructed under boulders and in rock crevices close to the shore; rabbit burrows and drainage holes are also much used. Only occasionally is the nest visible, and it may be up to 2 metres from the entrance.

Most of our breeding birds winter in sheltered waters close to the shore, making use of sea lochs and quiet harbours.

During early spring, 'gatherings' of black guillemots are in evidence on the sea close to their breeding sites. There is much excitement generated as pairs swim round each other, heads bobbing, tails cocked, and with bills open wide displaying that bright red gape. Incubation, a duty undertaken by both birds, often commences with the first egg; the second is laid up to three days later. It continues for 28 to 30 days; after another 35 to 40 days (and before they are fully fledged) the young vacate the nest and are soon to be heard giving voice to the high-pitched and wheezing call ('sphee-ee-ee-ee') we associate with this species.

Food is obtained in shallow waters, and now the black guillemot makes full use of its diving abilities, using wings and feet to great advantage as it hunts for small fish, in particular the butterfish. Crustaceans, marine worms and molluscs are also taken.

World distribution includes eastern North America from Maine northwards, and the archipelagos of Arctic Canada, Greenland, Iceland, the Faeroe Islands, Ireland, northern Britain, northern Denmark, Finland, Estonia, Norway, Sweden, the Kola Peninsula, and coastal regions of the White Sea.

Razorbill (*Alca torda*)

Length 41 cm (16 in)

Somewhat stouter than the guillemots but with the same short wings and contrasting plumage of black upperparts and white beneath. There is little doubt that the bird's common name is derived from the bill structure, which is compressed vertically, broad, grooved, and black, with a central white line crossing each mandible. A further white line extends from base of upper bill backwards to the eye. The secondaries are tipped with white, thus forming a narrow line along trailing edge of wing. The dark eye is deeply seated and often difficult to notice. Legs and feet are dark grey. From August through to March sides of head and chin become white, Young birds do not have the white bill markings, and some even retain the white on sides of head and chin into summer. Auks are often referred to as penguins of the northern hemisphere, a reference to their erect stance. Walking is difficult as they clumsily waddle about on their broad tarsi.

January and February see the early morning return of many razorbills to their sea-cliff sites, but they stay for only a few hours. During the next two months their numbers gradually increase, and the visits are now more frequent and for longer periods. Display involves both partners; it includes the mutual 'nibbling' of head and neck feathers; 'head-shaking' with partly opened bills revealing a yellow gape; and also what has been described as a 'butterfly-flight' when birds leave the ledges and fly for up to 60 seconds in 'slow motion' before reverting to their normal fast and direct flight on those short whirring wings. During this period of display they also engage in communal diving, and informal splashing-about or pivoting in the sea below the cliffs. The colony is altogether a very noisy place with that grating 'karrr, karrr' issuing from most of its inhabitants.

From early May onwards the single large pear-shaped egg is laid on a bare rock ledge; no nesting material is used. Although usually white, some are green or blue, all variably marked with browns and black. Incubation lasts from 34 to 39 days and is undertaken by both birds. The chick is cared for in the nest for another 18 days before being coaxed by its parents to flutter down to the sea below.

Razorbills dive for food from the sea surface; they are expert swimmers using feet and wings to propel them as they search for food. The diet includes small sprats, sand-eels, molluscs, and crustaceans.

An estimated breeding population for Britain and Ireland is put at 144,000 pairs.

There are three races: *A.t. torda* or nominate of Scandinavia and the Baltic Sea Islands; *A.t. pica*

Razorbill (*Alca torda*).

of N.E. Canada and N.E. United States; and *A.t. islandica*, found in W. Greenland, Iceland, Faeroe Islands, Britain, Ireland, the Channel Islands, and Brittany.

Birds from south-west of Britain and Ireland winter in the Bay of Biscay and the western Mediterranean. Those from southern Britain, including first-year non-breeders, travel to the North Sea. From Scotland large numbers travel to Scandinavian waters after breeding.

Cormorant (*Phalacrocorax carbo*)
Length 90 cm (35 in)

A large seabird with brown-black plumage, the cormorant is easily distinguished from the shag (*P, aristotelis*) by its large size (if a comparison is possible), but especially its white face and chin, and lack of an erect crest. At close quarters it may be possible to count the tail feathers which number 14, as against the shag's 12. During the breeding season it has white thigh patches, which become much more evident in flight. Often seen in an upright stance as it rests drying its outspread wings. Usually a silent bird but utters a variety of deep guttural calls at breeding colony, these include 'karrk' and 'kwarrk'.

Cormorants build bulky nests from material gathered in near vicinity; seaweed, bracken stems, and heather are mostly used. Usually an open coastal site is chosen, such as sea-stacks, broad cliff-ledges, sloping headlands, and small islands.

Cormorants (*Phalacrocorax carbo*) on the remains of an old jetty.

enclosed waters of Scottish sea lochs may be a little more vulnerable. At sea many types of flat fish are taken, and from fresh waters mainly trout and just a few young salmon.

The total breeding population of Britain and Ireland is in excess of 8,000 pairs, probably half of which occur in Scotland. In England the main colonies are in Northumberland, Isle of Wight, and from Dorset to the Scilly Isles. In Wales the strong-holds are Anglesey, Caernarvonshire, and Pembrokeshire. In Ireland they breed on practically all the coast. Seven races are recognised: the nominate *P.c. carbo* from E. Canada to the British Isles; *P.c. maroccanus*, N. Africa; *P.c. lugubus*, N.E. Africa; *P.c. sinensis*, C. Europe to India and China; *P.c. hanedae*, Japan; *P.c. novaehollandiae*, Australia and Tasmania; *P.c. steadi*, New Zealand and Chatham Island.

Chough (*Pyrrhocorax pyrrhocorax*)
Length 38 cm (15 in)

This was once a comparatively common British species but unfortunately its numbers have declined. Nowadays there is doubt if it still breeds anywhere in England (in a wild state); the last record of breeding seems to be around 1952 from Devon/Cornwall. Fortunately it still breeds in North Wales, the Isle of Man, Islay (off the S.W. coast of Scotland); its stronghold being the coastal areas between Cork and Donegal along western Ireland. The total British breeding population is probably less than 1,000 pairs.

The chough has the glossy black plumage typical of most members of the crow family; but that long decurved red bill and the red legs make it easy to identify. Sexes similar; the female's wing span is a little less than the male's.

Watch for the spectacular aerial displays around the sea-cliffs during the breeding season, April/May. A bulky nest is constructed, usually in a hole along the cliffs, or in a sea cave, and mostly inaccessible. Materials include twigs, heather stalks, roots, and other coarse vegetation, with a soft lining of sheep's wool and hair. On Islay many pairs choose to nest inside farm buildings or old deserted buildings. Inland cliffs, and disused copper and lead mines (North Wales) are also used. The three to six eggs are white spotted with brown and grey; they are incubated solely by the female for 17 or 18 days. A

Abroad, inland sites on marshes or in trees are also favoured. The male bird supplies most of the building material which the female fashions into a nest. Laying begins from early April onwards; the three to five elongated eggs are pale blue and covered in a chalky deposit, soon staining. They are laid at intervals of about two days and incubated by the parents in turn, for 28 to 30 days. The chicks are born blind, their eyes opening when about five days old. A regurgitated diet of fish is fed to them and after five weeks they are able to scramble about in and around the nest, but are unable to fly until seven or eight weeks old, and even then they will still be dependent on their parents for several more weeks.

Because of the cormorant's fish-eating habits they have, in the past, suffered large-scale persecution, when bounty schemes were introduced by some river authorities. However, there is little fear of fish stocks being affected in the open sea, but the diet chiefly of insects and small mammals is the norm; this may be supplemented with grain, fruit, and a variety of other vegetable matter. Both parents help feed the young (by regurgitation); they fledge at five-and-a-half weeks old. One of the chough's best-known calls is a rather musical and prolonged 'kyaaaa', another a high-pitched 'kwee-ow'. Altogether a noisy chattering bird both in flight and in gatherings on the ground.

There are eight races: *P.p. pyrrhocorax* of Scotland, Wales, Ireland and the Isle of Man; *P.p. erythrorhampus*, W. Europe; *P.p. barbarus*, Canary Isles, N.W. Africa; *P.p. baileyi*, N. Ethiopia; *P.p. docilis*, E. Europe to Arabia, Iran; *P.p. centralis*, C. Asia, Pakistan, N.W. India; *P.p. himalayanus*, N. India, Himalayas, W. China; and *P.p. brachypus*, N. China, N.E. Asia.

Peregrine Falcon (*Falco peregrinus*)
Length female 48 cm (19 in), male 38 cm (15 in)

This is the largest of the falcons breeding in Britain, where its population is put at a little over 1,000 pairs. It is probably only the remoteness and inaccessibility of many of its breeding sites that have helped the peregrine survive. The persistent stealing of eggs and young continues to jeopardise its breeding success, and even its survival, in parts of its range. Although regularly nesting on sea-cliff sites it is by no means confined to the coast. Crags, especially in mountainous terrain, are frequently used, as are treeless areas on moors, fells and mountainsides. A resident in Britain, it breeds in northern England, and sparingly in the south; also Wales; Scotland and Ireland.

The male (tiercel) has crown and sides of head blackish, white face with a blackish cheek patch, its slate-grey mantle contrasts with the buff underparts which are narrowly barred with black. Female (falcon) is larger than her mate, and quite often much darker. Juvenile birds are dark brown above, and buffish underparts are streaked, not barred.

When airborne those long pointed wings beat rapidly, and the flight is strong and powerful, interspersed with short glides. In search of prey, the peregrine towers high above its quarry, then suddenly and quite dramatically it swoops down at speeds well in excess of 240 kph (150 mph) to grab the victim, in mid-air, with its talons. It is said that over 120 different species of birds are preyed upon.

Outside the breeding season it is mainly a quiet bird, but should an intruder approach its nesting area then the peregrine becomes very noisy, producing a shrill chattering 'eee-yerrrk', or a long plaintive cry. The female's piercing scream once heard is never forgotten.

Most clutches are laid in May; the three or four whitish eggs with reddish-brown markings are deposited at two-day intervals in a scrape on some cliff ledge, no nest is built. The incubation takes 28 to 32 days and, although shared by both birds, the female undertakes the major share, thus allowing the male to provide the food. Small young are cared for and brooded by the female; as they grow in size so brooding becomes unnecessary and now both birds are able to go in search of food. After some five or six weeks the young are capable of flight, but they still remain within the nesting vicinity for a further short period.

The peregrine falcon is one of the world's most cosmopolitan breeding birds, with eighteen races extending it to all the continents; and the extreme forms differ considerably in size and depth of colour, but all have that distinctive white face with black cheek patches, and adults birds a buffish breast barred with black. *F.p. peali*, coast of W. Canada and W. USA; *F.p. anatum*, north C. America and South America; *F.p. cassini*, S. Chile, Tierra del Fuego, Falkland Islands; *F.p. peregrinus*, Europe to N. Russia and the Caucasus; *F.p. calidus*, N. Russia, N. Siberia to Southern Africa; *F.p. japonensis*, Siberia, Japan, Taiwan; *F.p. brookei*, Mediterranean, Asia Minor; *F.p. pelegrinoides*, North Africa, N. Sudan; *F.p. babylonicus*, N. India, Iraq to Mongolia; *F.p. peregrinator*, India, Sri Lanka to S. China; *F.p. minor*, Ghana to Ethiopia; *F.p. submelanogenys*, S.W. Australia; *F.p. macropus*, Australia (except S.W.); *F.p. madens*, Cape Verde Island; *F.p. radama*, Madagascar, Comoro Islands; *F.p. furuitii*, Volcano Islands; *F.p. ernesti*, Indonesia, Philippine Islands, New Guinea; *F.p. nesiotes*, New Hebrides, Loyalty Islands, New Caledonia.

Raven (*Corvus corax*)
Length 64 cm (25 in)

There is an estimated breeding population of 5,000 pairs in the British Isles. Even though there are many sea-cliff nesting sites in Britain, the raven is most certainly not confined to coastal regions; it seeks also the solitude of wild mountainous terrain. A comparatively common species in the Scottish Highlands and islands, and especially so in the Hebrides and Shetland; it still occurs regularly along the coasts of Cornwall, north Devon, Somerset, and the Isle of Wight; also Cumbria and the Isle of Man; with good numbers in Wales and Ireland.

The raven is a totally black bird, whose large body size (wingspan of male 140 cm, 55 in) and heavy bill help to distinguish it from the smaller (52 cm, 20 in, long) carrion crow (*Corvus corone*), coupled with the large wedge-shaped tail, and the deep oft-repeated call 'prruk, prruk'.

Often the same ledge and even the very same nest are used year after year, although they often have alternative sites within their territory which they use in successive years. Breeding birds are very territorial. Once having mated they probably stay together for life. Nesting can commence as early as February, with both birds gathering material to construct their bulky nest of twigs, heather stems and roots; the deep mud cup is lined with moss, wool and hair. Coastal nests are usually built on inaccessible sea-cliffs. During the mating season it is truly fascinating to watch their aerial displays as they wheel about, or soar high above and then tumble upside down.

The four to six bluish-white eggs, blotched and spotted with brown, are sometimes laid by the end of February, but peak laying is during March. Incubation continues for 21 to 22 days and is undertaken solely by the female, during which time she is fed by the male. Young ravens are born blind and naked, and for several days the female will brood them, after which food is collected by both parents and administered mainly by regurgitation. At six weeks or thereabouts, the young will leave the nest. The diet is extremely varied and includes all kinds of carrion, be it fish, animal, or bird; the raven also takes eggs and young of the larger ground-nesting birds. In some areas it has become dependent on the presence of sheep-farming, as the after-birth of lambing ewes is eagerly sought; insects and some vegetable matter are also taken.

Eight races extend it throughout most of the northern hemisphere: *C.c. principalis*, Alaska, Canada and N. United States; *C.c. sinuatua*, W.C. United States, Central America; *C.c. varius*, Iceland, Faeroe Islands; *C.c. corax*, Europe, W. Asia; *C.c. subcorax*, S.E. Europe, Asia Minor to

Pakistan; *C.c. tingitanus*, N. Africa; *C.c. tibetanus*, C. Asia, Himalayas; *C.c. kamtschaticus*, N.E. Asia, N. Japan.

Oystercatcher (*Haematopus ostralegus*)
Length 43 cm (17 in)

The oystercatcher's bold black and white plumage coupled with its ruby eye, that long bright orangey bill, and those pinkish legs, make identification a simple matter. In winter it acquires a white collar across the neck. Oystercatchers occur throughout the world and ten species have been recorded. It frequents rocky, shingly and sandy seashores, as well as maritime turf. Outside the breeding season it becomes very gregarious, with flocks comprising several hundred birds being quite common.

Food is taken either by probing the bill to seek out worms or insect larvae in mud and soil, or by attacking various molluscs such as cockles, limpets and mussels. It has been calculated that in winter a single bird consumes its own weight of wet shellfish each day. In South Wales during 1973–4 some 7,000 oystercatchers were culled because the extent of predation on the cockles was unacceptable to the cockle-gatherers whose livelihood was in jeopardy.

Oystercatcher chick hatching.

The breeding season extends from April/May to June when a simple scrape is made and lined with small pebbles, rabbit droppings, and tiny shells; a task shared by both sexes. Two to four pale buffish-brown eggs are laid; these are darkly blotched and spotted. Incubation is undertaken by both birds and continues for 24–27 days. The chicks stay in, or close to, the nest for one or two days and fledge at about five weeks old.

The flight is strong, fast and direct, with short and regular wing-beats, keeping low over the water or shore. A variety of calls have been recorded, the most familiar being the loud 'pic, pic, pic'. When alarmed it utters a strident 'kleep kleep'. A long piping trill, which varies in both volume and tempo, serves as the song.

In north-west Europe the oystercatcher is by no means confined as a breeding species to coastal

Oystercatcher (*Haematopus ostralegus*) on a wall.

regions. Recent years have seen it move inland up river valleys, and along the margins of lakes, sometimes even on moorland, and in open woodlands; particularly so in Scotland and northern England.

In Europe and Asia the oystercatcher breeds along all the coasts from the Kanin Peninsula on the Barents Sea to north-west France including most of the Baltic coast. In Iceland and the Faeroe Islands, *H.o. malacophaga*; the British Isles, *H.o. occidentalis*. In southern Europe its breeding range extends and is discontinuous from western Spain throughout the Mediterranean, the nominate *H.o. ostralegus*.

Common Eider (*Somateria mollissima*)
Length 58 cm (23 in)

Surely the best known of all the European sea ducks; certainly the largest, and by far the most numerous. The breeding population in Britain is estimated at 10,000 pairs; there are about 200,000 pairs in Iceland; and 300,000 pairs in the Baltic. The total European population is put at about two million individuals.

In breeding plumage the male is chiefly black and white in almost equal proportions. The sides of his crown are black, this extending to just below the eyes, and forwards narrowly towards the nostrils. A central white line extends from the crown to the pale green nape. Cheeks and throat are white; breast also white but suffused with pinkish-buff. Remainder of underparts are black, as are the rump, tail, flight feathers, and greater coverts. The females are boldly and entirely barred with light and dark brown. The males in eclipse plumage are mainly blackish-brown, with upper breast and area around neck boldly speckled white; wing coverts and tertiaries are white. First-year males are similar to adult males in eclipse, but have a whitish breast. Bill of both sexes is grey lightly tinged with green; legs and feet olive-green with blackish webs. The drake's best known call is a soft cooing 'a-oooo', another often heard is 'woo-hooo'.

As a rule the eider is a colonial nester either in a truly wild state or on the eider farms that exist in Iceland. The nest is a deep cup on mossy ground or amongst other suitable vegetation; this is copiously lined with the duck's own down. A normal clutch would be four to six large greenish eggs. Laying commences during late April in the more southerly parts of its range. Incubation continues for about 25 days and is undertaken solely by the female. She does not feed during this period so must build up her body weight in advance.

The blue mussel is the eider's principal food, but crustaceans are also taken; these are acquired by diving or dipping.

Outside the breeding season eiders are only to be found at sea, but usually close in to shore. Quite often flotillas of ducklings are to be seen, often in company with adult birds. At this time the males gather in large offshore flocks for the purpose of moulting.

Wintering eiders in European waters are often seen in large concentrations. Examples are: the southern Baltic and the Kattegat in the order of 750,000 individuals; in Icelandic waters 500,000. The wintering population of Great Britain is put at 50,000 to 60,000 birds. Some areas of note are: Scotland, Lothian Regions at Gullane Bay, and Leith-Musselburgh. Grampian in the Ythan estuary. England: Northumberland at Lindisfarne (Holy Island); Lincolnshire, the Wash; and Essex, the Colne estuary.

Six races are recognised: the Faeroe eider, *S.m. faeroensis* (restricted to the Faeroe Islands); the European eider, *S.m. mollissima*, breeds in Iceland, Denmark, Norway, Sweden, Finland, and Russia, south to Scotland and northern England, locally in Northern Ireland (especially offshore islets), the Netherlands, and Brittany. The northern eider, *S.m. borealis*, American eider, *S.m. dresseri*, and Pacific eider, *S.m. v-nigra*, extend the eider to the remainder of the Arctic, from Greenland eastwards to eastern Siberia.

Rock Pipit (*Anthus spinoletta*)
Length 16 cm (6¼ in) **Fig. 1(1)**

The largest of the British pipits. A slim bird; its olive-brown plumage slightly darker than, and not so heavily streaked with black as, the more familiar meadow pipit (*Anthus pratensis*). Look also for the dark legs, and smokey-grey feathers of outer tail; the legs of the meadow pipit are pale brown, and its outer tail feathers white.

This is the only small perching-bird confined to a life along the rocky shores of our coastline. Inland sightings are an exception and can be attributed to winter storms; for by choice the rock pipit is a coastal resident, and the British population is estimated at over 50,000 pairs. It can be seen along most of our rocky shores and islands, so this precludes extensive stretches of eastern, south-eastern and north-western coastlines. They are often foraging along the strandline of upper shores; listen for the call 'tsup' as they take to the wing when disturbed. Included in their diet are insects and their larvae, slugs, snails, worms, molluscs, and crustaceans, supplemented with flower seeds.

During April or early May the four to five eggs are laid in a well-hidden nest constructed by the female. It may be under vegetation growing between rocks a little above high-water mark; a cavity in a wall; on a steep embankment; or in a high rocky

Fig. 1 1. Rock pipit (*Anthus spinoletta*). 2. Shore larks (*Eremophila alpestris*). Birds in winter plumage.

crevice. The off-white eggs are densely speckled with brown and grey, and incubated by the female for 14 to 15 days. Feeding them is a task undertaken by both parents and beaksful of insects are provided for about two-and-a-half weeks, by which time the chicks will have fledged. Two broods are reared each season.

The world distribution includes nine races: *A.s. petrosus* of the British Isles; *A.s. rubescens*, N. and N.E. Asia, North America, Mexico; *A.s. pacificus*, W. Canada, W. USA, W. Mexico; *A.s. alticola*, S.W. USA, and N.W. Mexico; *A.s. japonicus*, E. Asia, Japan, E. China, Burma; *A.s. coutellii*, C. Asia, Tibet, W. China, N. India, Iran. *A.s. spinoletta*, S. and E. Europe; *A.s. kleinschmidti*, Faeroe Islands; and *A.s. littoralis*, N.W. Europe.

Arctic Skua (*Stercorarius parasiticus*)
Length 38 cm (15 in) (not including 7 cm, 2½ in, projection of central tail feathers)

Interbreeding of the two colour phases (light and dark) produces many intermediate forms. The dark phase is predominantly blackish-brown, a little less dark on underparts. The light phase has dark upperparts including nape, crown, and sides of upper face to just below the eye. Rest of plumage sullied white, greyish on upper breast. Legs and bill dark in both phases. In flight a pale patch on wings becomes apparent in both phases. The percentage of light-phase birds increases progressively northwards; for example approximately 35 per cent in Scotland and the Faeroe Islands, 40 per cent in northern Iceland, 90 per cent northern Norway, to almost 100 per cent in Spitsbergen and Greenland. Yet the small population in the Baltic area of Finland contains about 95 per cent dark-phase birds. The total breeding population in Britain is put at a little over 1,000 pairs. These are mainly confined to the north of Scotland, Orkney and Shetland; but small numbers do occur on the west coast of Scotland and the Outer Hebrides.

The arctic skua resembles a large gull, but has a much more rapid flight. It obtains much of its food by harassing other seabirds and causing them to disgorge or drop their food (especially kittiwakes and terns). It also kills small birds, and often takes eggs from the nests of terns and gulls.

Arctic skuas nest in loose colonies, and egg-laying begins from late May onwards. A simple scrape on moorland serves as a nest, into which two greenish or browny eggs are laid; these are blotched and spotted with dark brown. Both sexes share the incubation from three-and-a-half to four weeks, and the chicks are fed by regurgitation. At four weeks old they can fly, but it will be another four weeks before they become independent of their parents.

The arctic skua's breeding range extends to most areas around the northern hemisphere, especially between 60° and 70°N. Its breeding habitat in the main is coastal tundra or moorland. In winter it becomes oceanic, migrating southwards to between 30° and 50°S, including southern South America, South Africa and Australia, but does not quite venture into antarctic waters.

Throughout its range there are no sub-species.

Great Skua (*Catharacta skua*)
Length 58 cm (23 in)

The great skua, or 'bonxie' as it is known in Scotland, resembles a large gull; in size somewhere between the great and lesser black-backed gulls. Above it is dark brown streaked tawny but rather paler on the underside. In flight the great skua's broad and rounded wings exhibit a white patch at base of primaries. The body is much more stocky and powerful than a gull's; the dark brown bill shorter and thicker; whilst the tail and wings are relatively shorter. The legs are blackish. When displaying, it utters a slow, gruff laugh-like call 'hah-hah-hah', at other times a piercing 'skerr'.

Their moorland breeding areas are re-occupied during March/April and loose colonies formed. After a period of ceremonial bowing, and the parading of males with neck feathers fluffed-out, mating duly follows. The nest is a large cup-like depression; some are but sparsely lined, others are quite a substantial collection of dead moorland grasses. Two eggs comprise a standard clutch; these can vary from reddish-brown to pale buff, heavily blotched and spotted with dark browns. Incubation continues for 28 to 30 days and is shared by both sexes. The young are fed by regurgitation, and in our experience even newly-hatched chicks are called away from the nest to be fed. This outward trek for food and the return journey back to the nest presents an exhausting problem for the young chicks.

Food is obtained in several ways: by following trawler fleets to feed on disposed garbage and offal;

by harassing fish-eating birds such as gannets, so causing them to release their catch; by taking the eggs and young of other species; or by attacking and killing puffins as they leave their burrows.

Should a human intruder traverse one of the breeding areas, then he or she will be subjected to quite a ferocious aerial attack by successive parents. A frightening experience, especially if unexpected! The colonies are vacated during August when birds travel southwards to spend winter in the North Atlantic, with much smaller numbers visiting the western Mediterranean.

The great skua has the unique distinction of being the only bird to breed within both the Arctic and the Antarctic circles.

The nominate race, *C.s. skua*, breeds in Iceland, the Faeroe Islands, Shetland, Orkney, Outer Hebrides, and the extreme north of Scotland. The total British breeding population is about 4,000 birds, of which roughly half occur on the island of Foula which lies off the south-west coast of the Shetland mainland. Its complete range is N.W. Europe to E. Canada. Five sub-species complete its breeding range: *C.s. chilensis*, S. Chile to W. USA and E. Argentina; *C.s. antarctica*, Falkland Islands, Tristan da Cunha; *C.s. clarkei*, S. Georgia, S. Orkney Islands, S. Shetland Islands; *C.s. lonnbergi*, South Island of New Zealand to S. Australia; *C.s. intercedens*, Kerguelen Islands to S. Africa.

Storm Petrel (*Hydrobates pelagicus*)

Length 16.5 cm ($6\frac{1}{2}$ in)

Sometimes known as 'Mother Carey's chicken' (especially by sailors). This, the smallest of British seabirds, barely weighing 30 g (1 oz). It is almost entirely sooty-black including legs, bill and eyes; the hooked bill has long tubular nostrils. Look for the white rump, whitish area on underwing and, when in perfect condition, the narrow wing bar; also square-ended tail. Short bouts of gliding are often included in its otherwise 'fluttering' flight over the sea. In its search for food (which seems to be mainly zooplankton and small fish), it engages in brief spells of low hovering flight when its dangling feet 'patter' across the water surface.

Other than during the courtship period storm petrels are usually silent in flight; now a loud churring 'terr-chick' may be heard. A similar call is also uttered at the nest.

On land they manage to walk on tiptoe but only by rapidly fluttering their wings; at the nest-site they can only shuffle in and out very awkwardly with body resting on full length of tarsi. The hours of summer daylight are spent either out at sea or in the nesting chamber incubating the single egg or brooding their downy chick.

The peak return ashore for breeding purposes occurs in mid-May, although a small proportion of birds may precede this by up to a month. There are few mainland sites in Britain and Ireland; small and now-uninhabited islands are the strongholds. The nest is always well-hidden and a variety of differing sites are used; these include crevices in dry stone walls; unoccupied burrows of puffins, manx shearwaters or rabbits – often a side tunnel is excavated by the storm petrel itself if the soil is sufficiently loose; small side-chambers in sea-cliff caves; even deep under grassy turfs or stones. A shallow depression is fashioned and this is sometimes lined with a few small stones or a little dry vegetation. Normal time for egg-laying is during the last two weeks of June. The single roundish white egg has a few small red speckles; it is incubated for about 38 to 44 days, a task shared by both sexes for spells of two or three days' duration. Brooding of the downy silvery-grey chick is continuous for the first seven or eight days; from then on it spends the remainder of its nestling period alone, apart from a nightly visit with food by its parents. This is provided on a regular basis for about 50 days or so, and then irregularly for a further six or seven days. By now the young bird's body weight is almost 60 g (2 oz) (almost twice that of an adult), and most of the down will have been shed. At about this time the young bird's food supply is terminated by the parents. This encourages it to leave the nest and venture into the world outside; for the next few nights it practises 'wing-flapping' close to the entrance of the burrow. Eventually the time arrives when, unaided, it must make its way, nocturnally, towards the sea and launch itself into flight from a rock or other suitably prominent site.

It is almost certain that all the storm petrel's breeding colonies are not known, therefore to give a true account of its total population in Britain and Ireland is not possible. What can be said is that the two largest colonies, each in excess of 10,000 pairs, are on the islands of Inishvickillaun and Inishtearaght off Co. Kerry, Ireland. There are probably

60-plus known colonies, many of which are small; these extend from Shetland southwards along the west coast to the Isles of Scilly and the Channel Islands. Indications would suggest a decrease in the size of many of these colonies.

In terms of world distribution, the storm petrel is restricted as a breeding species to the eastern North Atlantic and the Mediterranean. Included in the former are: the Westmann Islands (Iceland); Lofoten Islands (Norway); the Faeroe Islands; Britain; Ireland; the Channel Islands; Brittany; Iberia; and the Canary Islands (occasionally). In the Mediterranean it breeds on islands off the Tunisian coast, and eastwards including Malta and Sicily.

One of its main wintering areas seems to be off the southern tip of South Africa.

Leach's Storm Petrel (*Oceanodrama leucorhoa*)
Length 20 cm (8 in)

A slightly larger but much rarer bird in Britain than the storm petrel, it weighs up to 60 g (2 oz). In general it has sooty-brown plumage, the upperparts having a bluish bloom, the underparts being somewhat paler. The white rump and forked tail are only good means of identification at close range. It has a pale band on wing coverts. The black hooked bill has a tubular nostril; legs and feet are black. Perhaps the best aids to identification are its long tern-like wings, and the constantly changing direction and speed of its buoyant flight.

Unlike the storm petrel, it has a variety of calls; when in flight close to its breeding colony, a guttural and variably-pitched purring 'wirra-wirra-wirra' interrupted at times by a harsh staccato 'wicka-wicka-wicka'. A series of long drawn-out crooning notes may be heard issuing from the nest burrows.

Leach's storm petrel will, if the soil is not too hard, excavate its own burrow, or it may choose a natural crevice in a cliff-face or among rocks. The nest is sometimes lined with sheep's wool or dry grasses. The single white egg is usually laid during May and is incubated by both parents for a period of seven weeks; the chick will remain in the nest for a further nine weeks.

Four colonies have been recorded in Britain and Ireland; these are on St Kilda, North Rona, Sula Sgeir, and the Flannan Islands. In all, several hundreds of pairs are thought to breed there, but this is purely an educated guess.

World distribution as a breeding species includes: Westmann Islands (Iceland); Lofoten Islands (Norway); the Faeroe Islands; off the Scottish coast (5 islands); also islands off western Ireland. On the North Atlantic coasts of North America some colonies on wooded islands are indeed vast; the population on Newfoundland is estimated in millions!

O.l. leucorhoa breeds N. Atlantic and N. Pacific; *O.l. beali*, S.E. Alaska to California. *O.l. kaedingi*, Guadelupe Island (Baja, California).

Wilson's Storm Petrel (*Oceanites oceanicus*)
Length 17.8 cm (7 in)

This is the other storm petrel, but is just a straggler to Britain and lucky to be seen. It has long legs with yellow feet that extend beyond the tail in flight. It breeds on sub-Antarctic islands.

List of Plant Species

Sea Spleenwort (*Asplenium marinum*)
Family Polypodiaceae (7) **Fig. 2(6)**

This handsome evergreen maritime fern, like all species that grow on rocks, has tough wiry stems; these penetrate into crevices holding the plant firmly in position. The rhizome (underground stem) is short and densely covered with blackish scales; the fronds often grow from it in tufts. These are usually 6 to 30 cm (2–12 in) long, lanceolate, and simply pinnate with up to 20 pinnae (leaves) on either side, each is about 1.5 to 4 cm ($\frac{1}{2}$–$1\frac{1}{2}$ in) long, ovate-oblong with a notched margin and of a leathery texture; the longest are towards the middle of the frond. Clusters of spore-bearing capsules (sori), each 3 to 5 mm long and covered by a brownish tissue (indusium), are produced on the reverse side of the leaves on the upper fork of the secondary veins and about halfway between the midvein and the leaf margin.

The spores ripen June to September and now the indusium bursts open to release the spores. Sea spleenwort grows in the crevices on sea-cliffs. Distribution is from the Isle of Wight to Cornwall; but most commonly on the west coast northwards to Shetland; then down the east coast to north Yorkshire; found on all coasts of Ireland.

Long-bracted Sedge (*Carex estensa*)
Family Cyperaceae (145) **Fig. 2(1)**

Occurs not only on damp coastal cliffs and rocks, but equally or probably more so on grassy salt marshes. A slender and somewhat rigid perennial growing to between 20 and 40 cm (8–16 in) tall. Stems are smooth, three-angled, as a rule are longer than the leaves, and sometimes curved. Leaves are very narrow, usually 2 mm wide or maybe less, with inrolled margin; they emanate more or less from base of stem. The single male spike is terminal, measures 10–15 × 1–2 mm, and is sessile; male glumes (small bracts with a flower in axil) are 3–3.5 mm, obvate and glossy-brown. Female spikes number two to four, these are touching, and measure 8–15 × 4–6 mm; the lower spikes have short peduncles (stalks). Female glumes are about 2 mm, broadly ovate and straw-coloured with patches of brown. The bracts are leaf-like, narrow, rigid, and many times longer than the spikes, and at length becoming more or less horizontal. It flowers from June to July. It can be found around coasts of British Isles north to Orkney, and is locally common.

Sea Pearlwort (*Sagina maritima*)
Family Caryophyllaceae (30) **Fig. 2(5)**

This small annual grows from 2 to 15 cm ($\frac{3}{4}$–6 in), has a slender taproot and its leaves often form a central rosette. The plant is void of hairs; has a longish central stem and many lateral ones which may be either of a prostrate or ascending habit, but the flower-stalks are always more or less erect. The small linear, fleshy leaves have a blunt point and are rounded at the back. Its flower has 4–0 tiny white petals and blunt concave sepals whose tips bend inwards. May to September is the flowering period. Also grows on dune slacks. Occurs locally all around the British coast.

Buck's-horn Plantain (*Plantago coronopus*)
Family Plantaginaceae (112) **Fig. 2(4)**

This has slender flower-spikes, and a rosette habit, typical of the plantains. The more or less prostrate

Fig. 2 1. Long-bracted sedge (*Carex extensa*). 2. Wild carrot (*Daucus carota* ssp. *gummifer*). 3. Carrot broomrape (*Orobanche maritima*). 4. Buck's-horn plantain (*Plantago coronopus*). 5. Sea pearlwort (*Sagina maritimum*). 6. Sea spleenwort (*Asplenium marinum*).

leaves are somewhat downy, about 2 to 6 cm ($\frac{3}{4}$–2$\frac{1}{2}$ in) long and very variable; they are narrow, linear, 1-nerved, and most often pinnatifid (deeply divided almost to base), occasionally toothed, and along with fruit (which is about 4 mm, 3–4 celled and 3–4 seeded) provide adequate characteristics for positive identification. It is very occasionally host to the carrot broomrape (*Orobanche maritima*). It flowers from May to July. As well as growing on sea-cliffs and in rock crevices it also occurs on sandy and gravelly soils. Can be found in most coastal areas throughout Britain and Ireland.

Wild Carrot (*Daucus carota* ssp. *gummifer*)
Family Umbelliferae (75) **Fig. 2(2)**
A fairly tall biennial, 30–100 cm (12–40 in) with only sparse foliage; stem is wiry, clothed with bristles, striate, and sparingly branched. It is readily distinguishable from other umbellifers by the foliage alone. The leaves are tripinnate (the primary and secondary divisions are themselves pinnate), segments narrow and acute; lower leaves are oblong with lanceolate leaflets; upper leaves larger and more triangular, leaf-stalk sheathed. Umbels are 3–4 cm (1–1$\frac{1}{2}$ in) in diameter and more or less flat with numerous rays; flowers are white, central flower in each umbel is usually red or purple; they are large in proportion to size and height of stem when compared with other umbellifers. Attracts numerous species of insects, and is host to the carrot broomrape (*Orobanche maritima*). The wild carrot ssp. *carota* occurs throughout the British Isles with the exception of eastern Sutherland and Shetland. However, we can only see the ssp. *gummifer* locally on sea-cliffs (and dunes) chiefly along the south coast. The ground lackey moth (*Malacosoma castrensis*) lays her eggs around its stems.

Carrot Broomrape (*Orobanche maritima*)
Family Orobanchaceae (107) **Fig. 2(3)**
Broomrapes are annual to perennial root-parasites, whose underground tubers attach themselves to the roots of a host plant, in this case the wild carrot (*Daucus carota* ssp. *gummifer*). Thus it is found only in those coastal regions that the wild carrot inhabits; very occasionally the buck's-horn plantain (*Plantago coronopus*) plays host. Erect, scaly flowering shoots, with terminal spikes, emanate from the

carrot broomrape's tubers; these are 10–50 cm (4–20 in) tall, purplish, more or less glabrous, and coated in a few brownish-purple scales below. The numerous flowers grow in a dense spike. The petals are pale sullied yellow with purple veins, and corollate; the extending upper lip is notched and curved forwards; the lower lip is three-lobed, the central lobe being largest, and serving as a landing place for visiting insects. The corolla measures 12–17 mm. Its bracts are purple, broad at base, becoming pointed, and more or less equal to length of flower. Is in bloom from June to July.

A rare plant, best known from Channel Isles, Kent, Dorset, Devon and Cornwall.

Sea Thrift (*Armeria maritima*)
Family Plumbaginaceae (95)
There is little doubt that this must be the best-known of all seaside plants. Not only does it occur on sea-cliffs and sea walls, but also in saltmarshes where it grows profusely. Flowering commences in late April and continues through to October in places; May to July is the best period. Suddenly thousands of little pink flower-heads begin to appear, each on top of its own stalk, and comprised of about 30 tiny individual flowers complete with 5 pink (occasionally white) petals, and a papery-textured tubular calyx 5 mm long and strengthened by five ribs which persists long after the flowers have died, hence the straw-coloured look after flowering. There is also an outer calyx to the entire head of flowers (cyme). Sea thrift is a perennial with all its leaves radical; they are long and grass-like but fleshy with a marginal fringe of hairs; they vary greatly in length depending on habitat. It occurs commonly throughout the British Isles. The caterpillar of the thrift clearwing moth (*Aegeria muscoeformis*) feeds on its roots, and that of the grass eggar moth (*Lasiocampa trifolii*) on its leaves. The Island of Jersey has its own species, Jersey thrift (*A. armeria*), which grows on stable sand dunes.

Sea Campion (*Silene maritima*)
Family Caryophyllaceae (30)
Sometimes known as seaside catchfly, this maritime perennial grows from 8 to 25 cm (3–10 in). It has numerous branched stems; some spreading and more or less prostrate forming a rosette of non-

flowering shoots. The other stems are erect, and each bears a terminal flower with five deeply-notched white petals, and a bladder-like calyx. The leaves are small at about 2 × 8 mm and glaucous. Flowering period is June to August. When growing in a sandy shingly situation, it is often in company with the yellow horned-poppy (*Glaucium flavum*) and Shore Centaury (*Centaurium littorale*). The larvae of the netted pug moth (*Eupithoecia venosata*) feed on its seed pods from late June to early August; larvae of Barrett's marbled coronet (*Hadena barrettii*) feed on its roots from June to August; those of the grey (*H. coesia*) feed on the buds, flowers and seeds; and of the pod lover (*H. lepida*) feed on its seeds late June to July, and again in September. Along the south coast the caterpillars of the feathered brindle moth (*Aporophyla australis*) feed on it, October to April. Sea campion is locally abundant all round the British Isles. Another campion, moss campion (*Silene acaulis*), which is really an arctic-alpine species with small pink flowers, can be found on the coastal cliffs of north-west Scotland.

Rock Samphire (*Crithmum maritimum*)
Family Umbelliferae (75)

A branched perennial which grows on the rocky south and west coasts of Britain, often in association with sea thrift (*Armeria maritima*) and scurvy-grass (*Cochlearia officinalis*). It grows 15–30 cm (6–12 in) and has solid, striate, and branching stems; the habit is compact, clustered, and sub-erect. The fleshy leaves are linear and divided several times, with leaflets on each side of a common stalk; leaf-stalks are short and stout with long membranous sheaths. It is in flower from July to September; the flowers are highly odorous, white and very small, about 2 mm across with minute pointed petals which turn inwards; they grow in umbels.

Rock samphire can be used in salads and as a pot herb. It occurs from Suffolk and Kent to Cornwall, then northwards to Ayr, and the isles of Lewis and Islay; also the coasts of Ireland where it is local in the north.

Rock Sea-lavender (*Limonium binervosum*)
Family Plumbaginaceae (95)

This is an herbaceous perennial with slender, wavy, flowering-stems of 5–30 cm (2–12 in); these are branched from below the middle to the base, and the branches repeatedly forked. Some of the lowest branches do not bear flowers. The flowering spikes are slender and erect; there are two rows of spikelets on the upperside of each spike. As a rule these are not set close together, therefore their outermost bracts do not overlap those of consecutive spikelets in the same row. Its flowers are corollate and 8 mm in diameter; the violet-blue petals overlap each other and are shallowly notched at the apex. The leaves are numerous and vary in length between 2 and 12.5 cm ($\frac{3}{4}$–5 in). They also range from lance-shaped to spoon-shaped but are always broadest above the middle. Look for flowers from July to September. Also grows on stabilised shingle. Occurs in Britain north to Lincolnshire and Wigtown.

Red Valerian (*Centranthus ruber*)
Family Valerianaceae (118)

An introduced, and somewhat glaucous, perennial of erect habit from 30 to 80 cm (12–31 in) tall, and often referred to as spur valerian. Although it inhabits coastal chalk cliffs where it can be locally abundant in suitable areas, it is by no means confined to the coast, and is frequently found on old walls and in waste places, as well as being a cultivated species. The stem is round in section and hollow; its radical leaves are about 10 cm (4 in) long, lance-shaped, and stalked; the upper ones are ovate to lance-shaped, sessile, smooth, and bluish-green. The flowers are hermaphrodite containing both functional-stamens, and an ovary; they are about 5 mm in diameter, usually red but occasionally white. They are arranged in long, terminal, panicled cymes; the narrow corolla-tube is 8–10 mm long; the slender spur is a little shorter but twice as long as the ovary. It flowers from June to August and occurs in suitable habitats throughout Britain and Ireland, particularly south-west England and south-east Ireland.

Rock samphire (*Crithmum maritimum*) and golden samphire (*Inula crithmoides*) on old red sandstone rock in Dyfed.

Scurvygrass (*Cochlearia officinalis* ssp. *officinalis*)
Family Cruciferae (21)

A salt-loving plant requiring a saline soil, biennial or perennial. It is not confined to sea-cliffs and banks, occurring also on saltmarshes. Varies in height from 5 to 50 cm (2–20 in), with many stems which are usually erect, but often stoloniferous and trailing. The leaves are markedly thick and fleshy, the radical ones stalked and somewhat kidney-shaped, rarely less than 1.5 cm ($\frac{1}{2}$ in) across; those on the stem are sessile and ovate, with a wavy, obtusely angular margin; they are broadly heart-shaped at base where they clasp the stem. The flowers are 8–10 mm in diameter, with white petals (rarely lilac) which are two to three times as long as sepals. It is in flower from May to August and widely distributed throughout the British Isles.

Other species include Danish scurvygrass *C. danica*, an annual with long-stemmed basal leaves which are about 1 cm ($\frac{1}{3}$ in) wide (narrower than *C. officinalis*), and heart-shaped at base; the stem-leaves are also mostly stalked, the lowest of which resemble ivy leaves with three to seven lobes; the uppermost leaves may be oblong-lanceolate. The flowers are only 4–5 mm in diameter with mauve or whitish petals which are barely twice as long as sepals. Locally common on sandy and rocky shores, banks and walls by the sea. In all British coastal areas. Flowers from January to June.

Scottish scurvygrass (*C. scotica*) is a biennial or annual which forms small compact tufts of 5–10 cm; the flowering stems are more or less prostrate. The basal leaves are long-stalked and grow in a rosette; they vary greatly in shape but have a heart-shaped base, are thick and fleshy and about 1 cm wide. Stem-leaves are either short-stalked or sessile but do not clasp stem. Flowers are 5–6 mm in diameter with pale mauve petals. Flowers from June to August. Occurs locally in coastal regions of N. Scotland, the Hebrides, Orkneys, Shetlands, and Isle of Man, also north and west Ireland.

Long-leaved scurvygrass (*C. anglica*) is a biennial or perennial with a slender tap-root and more or less erect flowering-stems of 8–35 cm (3–14 in). Basal leaves form a rosette, they are obovate and never heart-shaped at base. Stem-leaves are ovate or elliptical, may or may not be toothed, mainly sessile, but the upper leaves clasp the stem. Flowers 10–14 mm in diameter, and on average larger than the previous three species; they have white or pale mauve petals. April to June is the flowering period. With the exception of Orkney and Shetland it is locally common in estuaries and muddy shores all around the British Isles. All scurvygrasses are rich sources of ascorbic acid (vitamin C) and were much eaten in the past, especially so by sailors as a safeguard against, or a cure for, 'scurvy', a depraved state of the blood brought about by a diet void of fresh succulent vegetables and fruit.

Yellow-wort (*Blackstonia perfoliata*)
Family Gentianaceae (100)

Only in southern coastal regions can this pretty little annual be regarded as common. The cliffs of Dover are exceptionally good because the habitat of the yellow-wort is clay or calcareous banks and damp chalky places. It has a rosette habit and grows from 15 to 45 cm (6–18 in). The stems are simple and completely enveloped at the base by blunt, inversely egg-shaped leaves which are 1–2 cm ($\frac{1}{3}$–$\frac{3}{4}$ in) long, and narrow below; thus creating difficulties for creeping insects that would try to climb up the plant. Stem-leaves are smaller and broadly egg-shaped, set in distant pairs, and clasp the stem. Foliage and stems are bluish-green. The yellow flowers are corollate and 10–15 cm (4–6 in) in diameter with egg-shaped petals that terminate in a point; they are numerous and arranged in a forked panicle. Look for spiral twist of closed yellow buds. Calyx is divided to base into eight slender sepals. They are self-pollinated and in flower from June to October, and tend to open at their best on sunny days only. It is common in many southern coastal regions, but not so further north where it extends to Lancashire and Durham; it occurs in Kirkcudbright and Jersey; also from Meath to Sligo in S. Ireland.

Kidney Vetch (*Anthyllis vulneraria*)
Family Papilionaceae (51)

It likes shallow soils in dry places, and whilst distributed generally throughout the British Isles is more abundant near the sea and on calcareous soils. Another common name is 'ladies' fingers'. This tall, up to 60 cm (24 in), herbaceous perennial has a silky appearance, a woody rootstock and sub-erect stems. The leaves are pinnate and up to 14 cm ($5\frac{1}{2}$ in) long; leaflets of lower leaves are alternate, and vary in shape from elliptic to ovate, and acute or obtuse;

they are arranged on each side of a common stem, with a terminal one which is much the largest; the leaflets of upper leaves are opposite, and all similar, linear-oblong, and acute or with just a short narrow point. Foliage is bluish-white.

Flowers yellow or red; the head-like cyme is up to 4 cm (1½ in) across, and contains a cluster of flowers set in pairs. Each flower is tube-like about 12–15 mm long and on a short stalk, they can only be reached into by bees with their long proboscis. The calyx is somewhat woolly, more or less egg-shaped with pointed teeth, and contracted at the top. Petals exceed the calyx. Seed pods are about 3 mm long, stalked, and contain just one seed. The flowering period is June to September.

Included amongst the visiting bees are *Bombus silvarum*, *B. hortorum* and *B. muscorum*. It is one of the larval food plants of the mazarine blue butterfly (*Cyaniris semiargus*) which, although still common in other parts of Europe, is thought now to be extinct as a breeding-species in Britain. The small blue butterfly (*Cupido minimus*) lays her eggs singly on its flowers and later the caterpillar bores into the flowerhead feeding on the centre and the developing seed. In fact it is the only known foodplant of the small blue. It is also one of the foodplants of the chalk-hill blue butterfly (*Lysandra coridon*), and the caterpillars of the moths six-spot burnet (*Zygaena filipendulae*) and the six-belted clearwing (*Dipsosphecia scopigera*) feed on it.

Wood Vetch (*Vicia sylvatica*)
Family Papilionaceae (51)

Whilst the wood vetch grows on cliffs and shingle by the sea, its habitat includes thickets and rocky woodlands inland. This perennial has a trailing habit and a creeping rootstock. It grows up to 60 cm (24 in) long and the stem is furnished with forked tendrils. The stalkless leaflets number six to nine pairs, they are oblong-elliptic and from 5 to 20 mm long; they have semi-circular stipules (leaf-like appendages at base of leaflet), with bristle-like teeth at base. The blue or purple-veined white flowers are 15–20 mm in size and in loose, drooping racemes of 1 to 7 cm (⅓–2¾ in) each with 6 to 18 flowers. The stout flower-stalks are up to 10 cm (4 in) long, thus exceeding the leaves, the ultimate stalks being at least as long as the calyx-tube. Seed pods are oblong-lanceolate, pointed at either end, and about

30 mm long. The variety *condensata*, which is found on shingle banks, has few flowers and is densely leafy. Flowering period is June to August. A scattered distribution, occurring locally throughout the British Isles.

Cliff Sand-spurrey (*Spergularia rupicola*)
Family Caryophyllaceae (30)

A small perennial of prostrate habit having a branched, and relatively stoutish, woody stock. Its numerous creeping shoots tend to rise above the ground terminally, they are 5–15 cm (2–6 in) long, void of ridges or grooves, reddish-purple, furnished with glands, and densely hairy. The leaves are up to 15 mm long and about 2 mm wide; they are linear and fleshy with a horny tip, and with a silvery stipule (leaf-like appendage) at the base. The leaves of the lateral shoots are arranged in clusters at each node. The flowering-stalks may each bear up to 20 deep pink, 5-petalled flowers, each 8–10 mm across on a short (7–8 mm) hairy stalk. Sepals are 5 mm long equalling length of petals between which they alternate. June to September is the flowering period. Capsules are 5–7 mm long, seeds about 0.6 mm, black and wingless. Occurs on sea-cliffs, rocks and walls. Found locally and mainly on the coast between Hampshire and Cornwall, also northwards up to Ross on the west coast, including Isle of Man, Inner and Outer Hebrides. It is found on the east coast in Norfolk; Mid and East Lothian; and Aberdeen; also coasts of Ireland.

Wild Cabbage (*Brassica oleracea*)
Family Cruciferae (21)

A perennial or biennial plant with a more or less prostrate habit, but ends of stems tend to rise above ground. The stem is stout, twisted, and has leaf-scars below. Its hairless leaves are bluish-green, and upper surface is protected by waxy secretions. Lower leaves are stalked, wavy, and have a few small basal lobes. Upper leaves are stalkless, and may clasp the stem by up to one-third; they are oblong and do not broaden at the base. The flowers are borne on pedicels on either side of a common stem, the pedicels becoming progressively shorter upwards. The lemon yellow petals are 12–25 mm and twice as long as the erect sepals. The usually seedless beak is 5–10 mm long. Cells each have 8–

16 dark grey-brown seeds 2–4 mm in diameter. Flowers from May until August. Occurs on sea-cliffs along coasts of south and south-west England, also in Wales.

Black Mustard (*Brassica nigra*)
Family Cruciferae (21)

This herbaceous annual can be found growing wild on sea-cliffs, especially so in south-west England. If found inland it is fairly certain to be an escape. It has an erect habit with shoots of 30–110 cm (12–43 in) which are bristly below, becoming smooth and glaucous above; they have many ascending branches. All the leaves have stalks; the larger lower ones are shaped more or less like a lyre, they are rough, have a large terminal lobe, are grass-green, and up to 16 cm (6¼ in) long. The middle leaves have a wavy margin, and the uppermost are more or less lance-shaped, not toothed, void of hairs, and somewhat glaucous as already mentioned. Flowers are sub-corymbose, their stalks shorter than the calyx. Petals are bright yellow and about 8 mm long, twice the length of sepals. The fruits are 12–20 mm long, with short stalks of 2–3 mm long, they are held erect but always close to the stem. Beak slender 1.5 to 3 mm long, and always seedless. Cells each contain two to five dark reddish-brown seeds. Flowers from June to August.

Euphrasia (*Euphrasia occidentalis*)
Family Scrophulariaceae (106)

This maritime annual has an erect habit and the entire plant is downy. The stems are usually from 5 to 15 cm (2–6 in) long, purplish and robust, with either ascending or sub-erect, stout, basal branches. Because the leaves are all longer than the internodes, so the flower-spike becomes dense. The thick dull green leaves are 5–13 mm long with turned-down margins; they are covered with short whitish hairs, as well as equally short-stalked glands. Stem-leaves are obovate with apex somewhat obtuse, and with six to ten usually acute teeth; upper leaves are often larger than the lower ones. The white flowers are corollate, 5–7 mm across; upper lobes short, may be entire or notched; lower lip has longer lobes, they are broadish, and shallowly notched at apex, middle one is longest. Calyx has 8 to 14 acute teeth, and enlarges to a degree after flowering. In bloom

from May to August. Most common in south-west England, but can be found locally on grassy sea-cliffs around most of the British Isles. Other species of Euphrasia likely to be encountered include *E. foulaensis* which grows in coastal pasture as well as sea-cliffs in the Outer Hebrides, and from Moray to Shetland. It is an erect and rather slender annual of 2 to 8 cm (¾–3 in), whose reddish stem may be simple or with but a few branches. The thick, broad leaves are from 4 to 7 mm long, with few hairs.

The stem-leaves are few and distant, bluntly oval with four to six obtuse teeth. Corolla usually violet but occasionally white, and only 3 to 4 mm across; lobes of upper lip are entire, those of lower lip are longer and of more or less equal length; they are shallowly notched at apex. Calyx is mainly devoid of hairs, has teeth triangular to lance-shaped; it enlarges after flowering. In bloom July to August.

E. rotundifolia is an annual of 4–10 cm (1½–4 in) which grows very locally on sea-cliffs in the Outer Hebrides; also in Sutherland and Shetland. It is robust with an erect habit, often with a few very short branches emanating from halfway up the purplish stem. The rounded leaves are of almost equal length and breadth; they are 4–7 mm long, thick, dull green, and very hairy, especially on underside. The white corolla has a somewhat shaggy exterior, and measures about 6 mm; lobes of upper lip are short, and shallowly notched at apex; lobes of lower lip are a little longer, and also shallowly notched, middle lobe being expanded somewhat at apex. Calyx has short teeth, and becomes enlarged after flowering in July.

E. marshalii has stems of 5–15 cm (2–6 in) and grows on grassy sea-cliffs. It can be found locally in the Outer and Inner Hebrides, Mull of Galloway, also from Sutherland to Shetland. It is erect and robust as *E. rotundifolia* but has many branches which emanate from the base of its purple stem. It has larger leaves, 8–14 mm, which are covered quite densely with strong, wavy, white bristles; they are oval-shaped with a long wedge-shaped base, and are borne on upper part of stem. The corolla is 6–7 mm with a villous exterior, usually white, only rarely lilac or purple. Lobes of upper lip are short and slightly toothed; lower lip substantially longer than upper one, with lobes shallowly notched at apex; the middle lobe is longest and toothed at apex. Calyx has rather long teeth, and becomes enlarged after flowering in July and August.

Fennel (*Foeniculum vulgare*)
Family Umbelliferae (75)

An erect herbaceous perennial usually growing to heights of 60 to 130 cm (24–51 in). The stems are branched, round, polished and finely furrowed. They are solid, but develop a small hollow with age. Its dark green leaves have a sheath-like base where they emanate from the stem. They are much divided, 3–4 pinnate; the many slender, almost thread-like segments measure 1–5 cm, and are not all in the same plane. The small yellow flowers are 1–2 mm across and grow in terminal umbels of 4–8 cm ($1\frac{1}{2}$–3 in) in diameter; the umbels are composed of between 10 and 30 rays (stalks which together comprise the umbel), each bearing many flowers, and between 1 and 6 cm ($\frac{1}{3}$–$2\frac{1}{4}$ in) across. The entire plant has a strong aromatic odour. Young stems and leaves can be chopped and used raw in salads, or boiled and eaten as a vegetable. There is much doubt if it is truly indigenous, even in its coastal habitat; but inland it is almost certainly a garden escape. It flowers from July to October and occurs along the English and Welsh coasts southwards from Norfolk and Clwyd. Also in Ireland but mainly in the south and east.

Alexanders (*Smyrnium olusatrum*)
Family Umbelliferae (75)

A stout biennial which reaches heights of 50–150 cm (20–60 in) and is particularly common on waste places near the sea, but especially so on chalky sea-cliffs. Some good examples are the south-west coast of Anglesey, close to the sea in parts of Norfolk, and the sea-cliffs at and near to Dover; but Alexanders also grows near estuaries, and rivers which have a high salinity. The stem is solid (becoming hollow with age), round, furrowed, and branched (upper branches often opposite). Its dark green, shiny radical leaves are about 30 mm (12 in) long, and three-ternate (divided into three more or less equal parts, each leaflet again similarly divided a further twice); the stalked leaf-segments are ovate, and lobed or obtusely serrated. The leaves of upper stem are mostly opposite, and divided into three leaflets; base of leaf-stems are broadly sheathed. From late April until June, dense crowded clusters of greenish-yellow flowers, about 1.5 mm across, are borne on both terminal and lateral umbels. Alexanders is an introduced species which has been naturalised quite extensively in hedgerows, waste places and cliffs, especially in coastal areas. Whilst far commoner in the south its range does extend north to Dumbarton and Banff. It occurs throughout Ireland where it is most common near the sea, elsewhere inland very local.

Golden Samphire (*Inula crithmoides*)
Family Compositae (120)

Although this fleshy-stemmed perennial does occur on saltmarshes and shingle, it seems more truly at home on rocky maritime cliffs, where it withstands the regular sea-spray produced by high incoming tides. It is held securely in position by virtue of a long woody rootstock which gives rise to a cluster of erect smooth stems of 15–30 cm (16–12 in); these are furnished with numerous, narrow, fleshy, sessile leaves of 2.5–6 cm (1–$2\frac{1}{4}$ in), many of which have three teeth at apex. The flower-heads are few, about 2.5 cm (1 in) across and comprise an outer ring of golden-yellow ray florets, and an inner disc of small tubular orange-yellow florets. Flowering period is July to August. Golden samphire occurs locally in south-west England and Wales.

Tree Mallow (*Lavatera arborea*)
Family Malvaceae (37)

An almost tree-like downy perennial with stout, woody, erect stems of 60–300 cm (24–118 in) which reach up to 2.5 cm (1 in) in diameter. The leaves are stalked, velvety, pubescent, roundishly heart-shaped, have five to seven acutely scalloped lobes, and fan-like folds. Flowers borne terminally in simple or compound racemes; each is 3–4 cm ($1\frac{1}{4}$–$1\frac{1}{2}$ in) in diameter with broad spearhead-shaped, overlapping petals of pale rose-purple whose broad deep purple veins are united below. In flower July to September. A native on sea-cliffs and waste ground close to sea. Occurs from Sussex to Cornwall, and along west coasts up to Anglesey, Isle of Man, and Scotland to Ayr. Considered as an introduced species along east coast where it occurs up to Fife.

An Introduction to Seaweeds

For the purpose of generalisation it is true to say that green seaweeds (Chlorophyceae) occur principally on the upper shore and in pools. Brown seaweeds (Phaeophyceae) are found on the middle shore to the lower shore, and red seaweeds (Rhodophyceae) are mainly on the lower shore. There are of course many variations to this simple rule, depending on conditions locally, as will be seen within the descriptive text.

Most green seaweeds have thin delicate fronds and are seen to advantage only when under water.

The very large, tough, brown seaweeds of the genera *Laminaria*, *Saccorhiza*, and *Alaria* are collectively referred to as 'kelps'. The 'wracks' are those slippery, brown leathery weeds with strap-like fronds that emanate from an holdfast, and are usually dichotomously branched. There are of course many other brown seaweeds which are much smaller and rather delicate.

Red seaweeds are usually small to moderate in size, need very little light, and often grow under the shelter of large brown 'wracks' and 'kelps', or in the relative darkness of deep rock-pools. Most cannot survive if they dry out. Many other red seaweeds live in the 'kelp' forest of the sub-littoral zone, attached to the stalks of laminarians or other kelps. However, all of them can also be found growing on rocks in pools or in much deeper water. Those few red weeds that can survive periods of drying out often cover large areas on rocks and are exposed by receding tides. One red species which is able to survive in rock-pools higher up the shore is *Corallina officinalis*, the coral weed.

DESCRIPTION OF GENERAL STRUCTURAL OUTLINES
Figs. 3–5

1. Tubular frond (may or may not be inflated), unbranched.
2. Tubular branched fronds (may or may not be inflated).
3. Wavy membranous frond (with either a single or double layer of tissue).
4. Branched threads which are further irregularly branched and completely clothed in finer threads.
5. Branched threads which are further regularly plumosely branched.
6. Long unbranched threads.
7. Short unbranched threads.
8. Short branched threads.
9. Large, digitate, oar-shaped frond.
10. Long, wrinkled, undivided belt-like frond.
11. Frond dichotomously branched and dichotomously tipped.
12. Frond irregularly branched.
13. Twig-like frond with many branches.
14. Flat membranous frond, not branched.
15. Thin, hollow, membranous frond, not branched.
16. Branched stem with leaves laterally veined.
17. Ribbon-like frond, branched and with dichotomously tipped branchlets through which runs a divided midrib; no lateral veins.
18. Ribbon-like frond, branched with leaves, through which runs a divided midrib, no lateral veins.

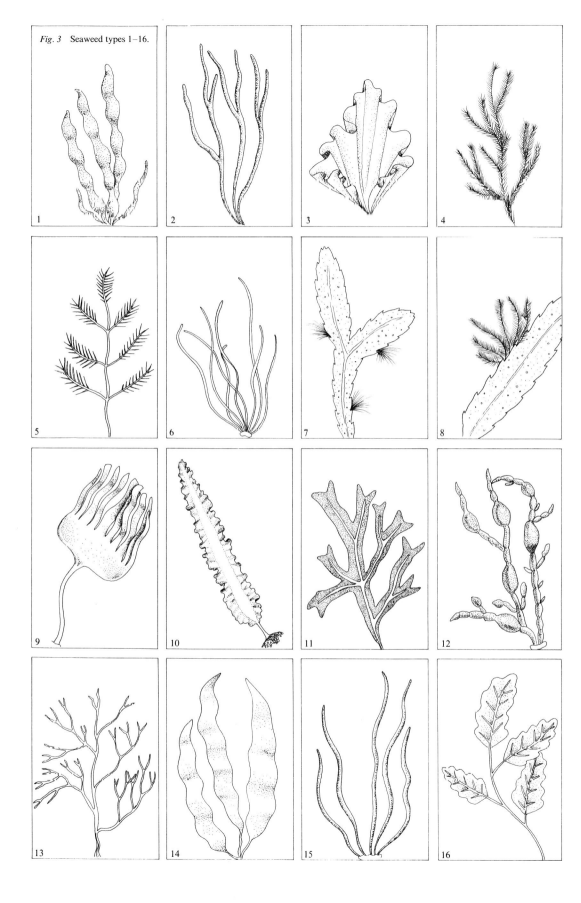

Fig. 3 Seaweed types 1–16.

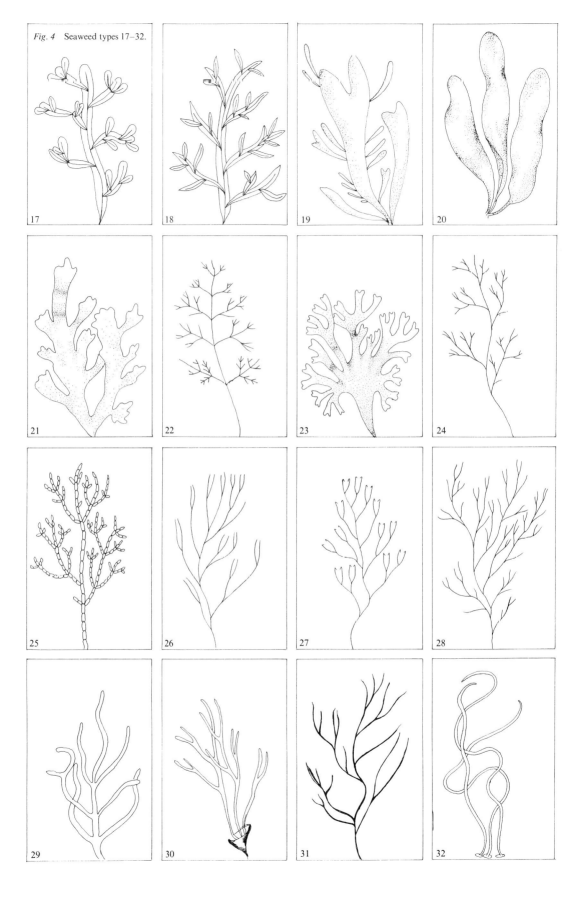

Fig. 4 Seaweed types 17–32.

Fig. 5   Seaweed types 33–50.

19. Somewhat palm-shaped frond of irregular outline with many marginal fingers.
20. Somewhat lobe-shaped frond.
21. Thin membranous frond, many-lobed, much divided, irregular outline.
22. Stem and branches either cylindrical or flat; branches more or less opposite are again similarly divided into branchlets.
23. Flat membranous frond with dichotomous stem and branches.
24. Stem and branches cylindrical; thread-like; or soft and limp. Branches and branchlets alternate, branchlets again divided into even smaller alternate branchlets.
25. Stem and branches constricted into bead-like sections.
26. Stem and branches thread-like, dichotomously branched, tips simple.
27. Stem and branches thread-like, dichotomously branched, tips divided into two inward-pointing hooks.
28. Thread-like stem with irregular, entangled, thread-like branches and branchlets.
29. Worm-like stem with similarly-shaped irregular branches.
30. Long and leathery branched straps emanating from a funnel-shaped disc.
31. Frond with irregular thread-like branches and branchlets.
32. Long, smooth and round 'bootlace'-type frond.
33. Frond dichotomous, circular in section, many-branched. Robust.
34. As 33 but less robust and with more subterminal branchlets.
35. Frond dichotomous, thin, flat and limp, with deep blunt-ended tips.
36. Frond flat with prominent midrib, branches irregular and dichotomous. Frond edges serrated.
37. Frond more or less flat, often wavy-edged. Prominent midrib with air bladders on either side.
38. Frond irregularly branched, continually varying in thickness.
39. Frond limp; branches irregular and covered in short fine hairs.
40. A moss-like growth; branches numerous and irregular resembling a string of small oval beads on *close* inspection.
41. Branches exactly opposite. Main stem and branches are a series of tiny oval, flattened, bead-like sections on *close* inspection.
42. Branches, branchlets and further sub-divisions are flat and alternate. Branches become progressively shorter towards tip of main stem, as do the branchlets.
43. Flat, ribbon-like and irregular dichotomous branches with dichotomously-tipped branchlets.
44. Irregularly branched with feather-like branches and branchlets.
45. Stem and main branches are comparatively thick. Many branchlets and further sub-divisions.
46. Fronds short-stemmed, elongated and leaf-like. Margin wavy but undivided.
47. Has a long and continuous stalk from one side of which emanate slender branches; these terminate in membranous dichotomies, thus producing a fan-like effect.
48. Stem and branches well developed and somewhat wavy. Branches irregular but main ones are often alternate. Branchlets occur approximately in sequences of four, first on one side and then the other.
49. Main stem and branches are flattened. Entire plant appears plumose.
50. Main stem irregularly-branched. Branches produce numerous upward-curving branchlets on *one side only* giving an almost comb-like appearance.

SEAWEEDS FOUND ON SALTMARSHES

Bostrychia scorpioides
Family Rhodophyceae **Fig. 4(28)**

A small, up to 10 cm (4 in) widely-distributed species which may be locally abundant. It has no holdfast and grows among the lower stems of several saltmarsh plants including common seablite (*Suaeda maritima*) and sea purslane (*Halimione portulacoides*). The plant is stiff and wiry with entangled, irregular, thread-like branches and branchlets. This red seaweed grows on the edge of saltmarshes. A similar-looking, but almost black, species *Ahnfeltia plicata* occurs on middle shore in pools on exposed beaches.

SEAWEEDS FOUND ON ALL SHORE ZONES

Enteromorpha compressa
Family Chlorophyceae **Fig. 3(2)** very common/common

This is just one of several *Enteromorpha* species which answer to the following description. It has long and narrow, inflated or compressed tubular green fronds of up to 60 cm (24 in) which emanate from a disc-type holdfast; they are branched once and may even have branchlets. The branches are blunt at tip and are narrowest where they join the main frond. Occurs on stones, rocks, piers, mudbanks in estuaries, in pools and in brackish water. Often grows prolifically forming a carpet of green.

Enteromorpha intestinalis
Family Chlorophyceae **Fig. 3(1)** very common/common

There are six common species of *Enteromorpha* which like *E. intestinalis* have long narrow, inflated, tubular, unbranched, mid to bright green fronds. These are usually slightly constricted at irregular intervals and up to 60 cm (24 in) long by up to 7 cm ($2\frac{3}{4}$ in) wide (broader than *E. compressa*). The frond ends are blunt; the short stalk is attached by a disc to rocks, stones and shells. Often abundant in muddy estuaries and brackish pools; by the summer months may carpet wide areas. When the plant becomes older, pieces begin to break off and lengths may be found washed up along drift-line or seen floating around in the sea.

Enteromorpha linza
Family Chlorophyceae **Fig.3 (1)** very common/common

This has long, narrow, bright green fronds, *not* inflated. They are usually wavy and up to 50 cm (20 in) long with a short hollow stalk; sometimes spirally twisted and quite ribbon-like. Attached to stones, shells and other seaweeds by a small disc. Often grows alongside *Ulva lactuca*, sea lettuce, with which it might be confused. Also occurs in pools and close to out-flows of fresh water.

Ulva lactuca (Sea Lettuce)
Family Chlorophyceae **Fig. 3(3)** Very common/common

This has very thin membranous, irregularly-shaped fronds which, when in water, open out to look very much like lettuce leaves. Pale green when young, shading through bright green when mature to dark green in age. When spores are released the margins often look whitish. It grows in bunches and is usually up to 45 cm (18 in) long; it has a small stalk. Found on rocks and mudflats especially in sheltered bays and often very prolific where sewage outlet feeds into sea; most abundant during July and August. This worldwide species occurs on all shore zones but not on upper beach except when cast up on shore.

Ectocarpus viridis
Family Phaeophyceae **Fig. 3(4)** Very common/common

Produces limp yellowish-brown to olive-brown fronds comprised of branched threads which are further irregularly divided and sub-divided into hair-like filaments; up to 25 cm (10 in) long. There are several related genera which together have more than 40 similar species; positive identification can only be ascertained after examination under microscope. Grows on rocks, shells, and the fronds of larger seaweeds.

Porphyra umbilicalis (Purple Laver)
Family Phaeophyceae **Fig. 3(3)** Very common/common

Green when young and similar to *Ulva lactuca*, sea lettuce, but has double layer of tissue, darkens with age to brownish-purple, and finally black. If exposed to sunlight for long periods, becomes somewhat bleached and during winter months looks very much like a piece of black plastic. The fronds are of no definite shape and up to 10 cm (4 in) in diameter, usually very irregular and often deeply lobed. Habitat is exposed beaches on rocks, stones and shells (when these become sand-covered the fronds appear to be growing from the sand). There are three species of *Porphyra*, all very similar, and gathered in parts of South Wales, especially Swansea and Llanelly, to make that local delicacy 'laverbread'.

SEAWEEDS FOUND ON UPPER SHORE ZONE

Enteromorpha clathrata
Family Chlorophyceae **Fig. 3(15)** Very common/ common

A rather delicate species with long narrow, un-branched, tubular, mid-green fronds; only 15 mm in diameter and up to 25 cm (10 in) long. Attached by a disc holdfast to stones, shells and other seaweeds in pools or shallow water. Also occurs on middle shore.

Chaetomorpha aerea
Family Chlorophyceae **Fig. 3(6)** Very common/ common

This has numerous long, tangled, unbranched, mid-green threads which emanate from a source with no obvious holdfast. Often entangled around other species of seaweed. Occurs in brackish highwater pools; also on estuarine mudbanks. It is most apparent during the summer months. Also found on middle shore zone.

Fucus spiralis (Flat Wrack or Twisted Wrack)
Family Phaeophyceae **Fig. 3(11)** Very common/ common

This has broad, moist, and more or less flat, olive brown fronds; greenish-black when dry. The smooth-edged fronds are dichotomously branched, up to 38 cm (15 in) long and have prominent midrib. Branches are partially spiralled and their bluntly-forked tips are swollen when bearing the granular fruiting-bodies, these do not extend to the trans-lucent rim. Does not have air-bladders. Also occurs on middle shore. Grows on rocks usually above the beds of *F. vesiculosus*, bladder wrack, and just below *Pelvetia canaliculata*. It is not found on rocky shores which are extremely exposed.

Pelvetia canaliculata (Channelled Wrack)
Family Phaeophyceae **Fig. 3(11)** Very common/ common

This has a tufted growth and olive brown dichoto-mously-branched fronds of up to 15 cm (6 in) long, about 8 mm wide with one margin inrolled which

forms a channel lengthwise. The fronds terminate in bluntly-forked tips which are often swollen and granular. It has a disc holdfast and grows on rocks from splash zone to upper shore zone reaching the beds of *Fucus spiralis*, flat wrack. Becomes blackish and very brittle if exposed for long periods. Also occurs in estuaries and saltmarshes.

SEAWEEDS FOUND ON MIDDLE SHORE ZONE

Bryopsis plumosa
Family Chlorophyceae **Fig. 3(5)** Very common/ common

A rather shiny, limp, and branched plant; the branches becoming shorter towards the top; each branch has an upper and lower row of numerous fine branchlets which are produced along the outer two-thirds of the branch giving a feather-like ap-pearance. A dark green seaweed up to 10 cm (4 in) high which grows in pools, under ledges and on rocks; found especially in steep-sided pools.

Cladophora rupestris
Family Chlorophyceae **Fig. 3(4)** Very common/ common

Forms dark green bushy tufts of up to 15 cm (6 in), which not only look but feel coarse and wiry. The main stem produces branches which in turn give rise to secondary branches; all of which, including main stem, produce thick thread-like branchlets. Grows on shady rocks in damp situations, and especially among and under the larger species of brown seaweeds, particularly *Ascophyllum nodo-sum*, knotted wrack, and *Fucus vesiculosus*, bladder wrack. A very conspicuous species in summer when it is well developed, but dies back in winter months. It also occurs in shallow pools of low salinity, and on estuarine mudbanks.

Codium fragile
Family Chlorophyceae **Fig. 5(33)** Very common/ common

Produces dichotomous dark green branches which are circular in section and have a softish, almost sponge-like, texture. Many-branched and grows up to 45 cm (18 in). Roots of holdfast compact as if united. Forms dense clumps in sheltered areas along rocky coasts.

Codium tomentosum
Family Chlorophyceae **Fig. 6(31)** Very common/common

Very much like *Codium fragile* but less robust, with more small, sub-terminal branchlets. Dark green with a felt-like texture, branches are circular in section; up to 30 cm (12 in) in length; at its longest during winter months. Roots of holdfast so closely woven they appear to be all in one piece. Grows in deep shady pools along sheltered coasts. Also occurs on lower shore.

Enteromorpha clathrata
Described in upper shore section.

Chaetomorpha aerea
Described in upper shore section.

Ascophyllum nodosum (Egg or Knotted Wrack)
Family Phaeophyceae **Fig. 3(12)** Very common/common

Olive-green, irregularly-branched, up to 90 cm (35 in) long and with a disc holdfast. Fronds are leathery, slender, with no midribs, and produce egg-shaped bladders singly and at intervals down middle. During spring and early summer yellowish-green, raisin-like fruiting-bodies are produced on short stalks along frond margin; these are often in pairs and opposite. Plant becomes greenish-black if conditions permit it to dry out. Occurs on rocky coasts where, if reasonably sheltered, its growth is so abundant as to carpet large areas. Whilst on the more exposed rocky coasts, strong wave action quickly breaks up the plant leaving only short frondal remains. Usually occurs above the beds of *Fucus serratus*, saw wrack, but is more closely associated with *Fucus spiralis*, flat wrack.

Aspercoccus fistulosus
Family Phaeophyceae **Fig. 3(15)** Very common/common

Several hollow, unbranched, membranous fronds emanate from a disc holdfast. When young these are green, but soon become olive-brown and grow up to 60 cm (24 in). Fronds are about 1 cm ($\frac{1}{3}$ in) wide and somewhat shallowly constricted at intervals along length. During reproductive season it becomes stippled with very small spore cells which give rise to a fine filamentous growth, thus producing an overall whiskery appearance. Occurs on stones and pebbles, also in shallow pools. Particularly common during summer months.

Chordaria flagelliformis
Family Phaeophyceae **Fig. 4(31)** Very common/common

A dark brown frond. Central stalk has many long, thin, curving branches each of which produces several fine branchlets. Entire frond is gelatinous, consisting of strings of a stiff jelly-like substance. Occurs on rocks and in pools from July to November.

Cladostephus verticillatus
Family Phaeophyceae **Fig. 4(31)** Very common/common

Stiff, dark greenish-brown, and grows up to 15 cm (6 in). The entire plant is clothed in small whorls of tiny hooks, each whorl being quite distinguishable. Of wide distribution; grows on rocks and other seaweeds.

Cladostephus spongiosus
Family Phaeophyceae **Fig. 4(31)** Very common/common

Stiff, dark greenish-brown, and grows up to 7 cm ($2\frac{3}{4}$ in). Very similar in many respects to *C. verticillatus*, but the small whorls of tiny hooks around stems and branches appear at first to coalesce, and individual whorls only become obvious after close inspection. Grows on rocks and is widely distributed.

Desmotrichum undulatum

Family Phaeophyceae **Fig. 3(14)** Very common/ common

A flat, very thin, membranous, unbranched, olive-brown frond up to 30 cm (12 in) long, only 1.25 cm (½ in) wide, with a short bristly stalk emanating from a small disc holdfast. Grows on other seaweeds, for example *Chorda filum*, bootlace weed. A widely-distributed species which attains maximum length during summer and dies back as autumn approaches.

Dictyota dichotoma

Family Phaeophyceae **Fig. 5(35)** Very common/ common

A thin, flat and limp, dichotomous, olive brown frond which pales towards the deep blunt-edged tips. Has several fine longitudinal lines, is slightly spotted, and grows up to 30 cm (12 in). Emanates from its disc holdfast on rocks and small stones. Rather abundant on south and west coasts, but is far from common in north and east.

Saw or serrated wrack (*Fucus serratus*).

Elachista fucicola

Family Phaeophyceae **Fig. 3(7)** Very common/ common

Could be described as resembling a small shaving brush. A greenish-brown weed which consists of numerous simple, radiating, stiffish threads each about 2.5 cm (1 in) in length, which grow from the margins of large seaweeds such as *Fucus vesiculosus*, bladder wrack, and *F. serratus*, saw wrack. Widespread distribution and at times quite abundant.

Fucus ceranoides (Horned Wrack)

Family Phaeophyceae **Fig. 3(11)** Very common/ common

An olive-brown, dichotomously-branched weed which becomes greenish-black when dry. The fronds, which reach up to 90 cm (35 in) in length, have a disc holdfast, they are thin, more or less flat and have a well-defined midrib. Margins are entire; the tips forked and pointed. Often has additional small lateral, fan-shaped, pointed fronds. It is void of bladders. Grows on rocks and stones on middle shore; also found in brackish waters especially at the mouths of rivers and estuaries.

Fucus serratus (Serrated Wrack or Saw Wrack)

Family Phaeophyceae **Fig. 5(36)** Very common/ common

Olive-brown frond up to 90 cm (35 in) long which becomes greenish-black when dry. It is smooth, and flat with a prominent midrib; the branches are irregular and dichotomous. Frond edges are serrated; forked tips are bluntly rounded, and during autumn and winter they bear flat pointed fruiting-bodies which are also serrated and up to 5 cm (2 in) long. Tips of infertile fronds are often furnished with a scattering of silky hairs. There are no bladders present. Grows on rocks and stones, on all but the most exposed coasts. Occurs on the lower middle shore zone often in association with *Fucus vesiculosus*, bladder wrack, and *Ascophyllum nodosum*, egg wrack; also found on lower shore zone.

Fucus spiralis (Flat Wrack or Twisted Wrack)

Described in upper shore section.

Fucus vesiculosus (Bladder Wrack or Cutweed)

Family Phaeophyceae **Fig. 5(37)** Very common/common

Olive-brown fronds of up to 90 cm (35 in) which, like other *Fucus* species, become greenish-black when dry. Fronds more or less flat and dichotomously branched, often wavy-edged. Prominent midrib with air-bladders at intervals on either side; these may be arranged in opposite pairs. Swollen, glandular fruiting-bodies are borne at the ends of the lateral branches, and these are sometimes double pointed. Unlike *F. spiralis*, flat wrack, the rims of forked tips are not translucent. Often the older plants have more numerous but smaller air-bladders. Where the weed grows in areas subjected to heavy surf action, the bladders may not develop at all, and the lower part of frond could be worn away leaving only the tough cord-like midrib attached to the stones and rocks. Also occurs in brackish waters, creeks, and estuaries. Very widespread and abundant.

Sea belt (*Laminaria saccharina*).

Laminaria saccharina (Sea Belt or Sugary Wrack)

Family Phaeophyceae **Fig. 3(10)** Very common/common

Long, undivided, yellowish-brown frond with a short thick stalk, and a strong root-like holdfast attached to small rocks or stones. The frond has no midrib, is of leathery texture with a wavy margin and markedly wrinkled along its entire length. Usually grows up to 3 m (10 ft) in length and up to 30 cm (12 in) wide; occasionally much longer. Severe wave action often causes damage to fronds and pieces may be cast up along drift-line. New fronds grow each year from end of stalk. Occurs on stones in larger pools, and on muddy sand-flats; may carpet extensive areas at sea's edge. Also found on lower shore and sub-littoral zones. It is of wide distribution.

Leathesia difformis

Family Phaeophyceae Very common/common

This can only be described as a roundish, gelatinous, somewhat lobed, lump; olive-brown in colour and up to 5 cm (2 in) in diameter. Solid when small, becoming hollow as size increases. Perhaps more usually found attached to the fronds of other seaweeds. *Leathesia difformis* also occurs on rocks

and stones. First appears about March, becoming abundant by July. It has usually been washed away from its 'host' as November approaches. Also found on lower shore.

Mesogloia vermiculata

Family Phaeophyceae **Fig. 5(38)** Very common/common

Frond yellowish-brown, up to 7 cm tall and irregularly branched. Branches and branchlets are slimy and solid; a special feature is their continual varying thickness. This widely-distributed species shows a preference for the more sheltered stretches of shore, where it grows on stones and rocks, in the deeper pools. It is most abundant along the northeast coastline but common almost everywhere. Also found on lower shore.

Punctaria latifolia

Family Phaeophyceae **Fig. 3(14)** Very common/common

Fronds thin and membranous, up to 30 cm (12 in) long and up to 8 cm (3 in) wide with *round* spots. The lance-shaped fronds are unbranched and have wavy margins. Wider than *P. plantaginea*, whose fronds are covered in *oval* spots and which also occurs on lower shore. Grows in shallow pools where it is attached by a small disc holdfast to rocks; stalk very short. Occurs rather high on middle shore and dies back towards early autumn.

Punctaria plantaginea

Family Phaeophyceae **Fig. 3(14)** Very common/common

Very similar to the previous species, *P. latifolia*, but the thin wavy membranous fronds are dotted all over with *oval* spots, are narrower, up to 5 cm (2 in) wide, and terminate in a distinct point. Attains maximum size in summer months when it can be found attached, by a small disc holdfast, to rocks and other species of seaweeds. Occurs in pools on middle and lower shores. Usually dies back before autumn.

Sauvageaugloia griffithsiana

Family Phaeophyceae **Fig. 5(39)** Very common/common

This has slimy and somewhat limp olive brown fronds, which are more usually between 10 and 20 cm (4–8 in) but may on occasions be up to 30 cm (12 in) or even longer. The branches are irregular, covered with short fine hairs, and may or may not be partially hollow. Grows on rocks and stones in pools low on middle shore. Widely distributed around Britain's coasts.

Scytosiphon lomentaria

Family Phaeophyceae **Fig. 3(15)** Very common/common

Unlike *Asperococcus fistulosa*, whose tubular fronds are only very slightly constricted at irregular intervals, those of *Scytosiphon lomentaria* are sharply and regularly constricted to produce an effect not unlike a string of miniature sausages. Several olive brown fronds arise from a single disc holdfast which may be attached to rocks and stones in pools, or limpet shells, seaweeds and even eel-grass (*Zostera*). The tubular fronds are up to 45 cm (18 in) long, shiny and unbranched. Widely distributed, *Scytosiphon* is probably the most common of our larger brown seaweeds.

Sphacellaria cirrhosa

Family Phaeophyceae **Fig. 3(8)** Very common/common

The frond is composed of stiff and (but not obviously) branched threads of up to 2.5 cm (1 in). These form small dense brown tufts which are to be found growing on the fronds of larger weeds. The exceedingly small, opposite branchlets are only evident with close inspection.

Ectocarpus confervoides

Family Phaeophyceae **Fig. 3(4)** Fairly common

The genus *Ectocarpus* contains 20 or so species, and, coupled with the 20 other very similar species in related genera, positive identification, if necessary, can only be determined by an expert after painstaking work with a microscope. Therefore the following description can only generalise. The thread-like fronds are some shade of pale greenish-brown and up to 30 cm (12 in) long. The plant is limp and has thread-like branches which are themselves further divided. When submerged in water, being so fine, it sways about; out of water it hangs like wet matted hair. Fronds very variable and gelatinous. More usually grows on other species of seaweeds, but can also be found on rocks. Occurs on both the middle and lower shores.

Himanthalia elongata (Thong Weed)

Family Phaeophyceae **Fig. 4(30)** Fairly common

This olive brown, long, leathery, strap-shaped weed usually grows to about 120 cm (47 in) long, but may be a good deal longer. The fronds are dichotomous and emanate from a funnel-shaped base which is about 4 cm (1½ in) high and has itself developed from a button-like growth. The spotted parts of the fronds indicate the reproductive areas. Occurs on rocks along exposed coasts; often the thong-like frond becomes detached from the funnel-shaped base which has been likened to the chanterelle mushroom (*Cantharellus cibarius*). Widely distributed and locally abundant; also found on lower shore.

Ahnfeltia plicata

Family Rhodophyceae **Fig. 4(28)** Very common/common

An entangled growth of stiff wiry texture. Very dark brownish-red fronds with very irregular thread-like branches which are themselves sub-divided. A densely-tufted plant up to 30 cm (12 in) tall and 10 cm (4 in) across, which appears to be black unless closely inspected. Grows on rocks and stones in pools, especially those on the more exposed beaches. Also occurs on lower shore.

Catenella repens

Family Rhodophyceae **Fig. 5(40)** Very common/ common

Plant is only 1.2 cm ($\frac{1}{2}$ in) high and looks like a dark purple or almost black moss-like growth in sheltered cracks amongst rocks. The branches are numerous, irregular and jointed; only on close inspection will they be seen to resemble a string of very small oval-shaped beads. Grows in and amongst the holdfasts of larger seaweeds such as *Pelvetia canaliculata*, channelled wrack, and *Fucus spiralis*, flat wrack. Often covers substantial areas on steep sheltered rock faces. Holdfast consists of an interlaced system of threads. Occurs on upper middle zone.

Ceramium acanthonotum

Family Rhodophyceae **Fig. 4(27)** Very common/ common

Frond is a system of dichotomous threads which terminate in two inward-curved hooks (when viewed under a hand lens they are seen to resemble very tiny claws). Each thread-like branch and branchlet is banded alternately silver and pink. Grows up to 10 cm (4 in) tall and occurs on rocks which are subjected to wave action. Also found on lower shore. There are other species of *Ceramium* which might fit this description.

Ceramium rubrum

Family Rhodophyceae **Fig. 4(27)** Very common/ common

Frond similar in construction to *C. acanthonotum* with terminal hooked inward-curved tips, but grows in tufts of up to 30 cm (12 in). Varies in colour from brownish-red to deep red, and is alternately banded in light and dark sections. Seen underwater in bright sunlight, it often appears greenish-yellow.

Grows commonly on rocks, debris, and other seaweeds; also found in pools where it produces a tufted growth.

Chondrus crispus (Carragheen)

Family Rhodophyceae **Fig. 4(23)** Very common/ common

Although more usually some shade of reddish brown, the tough flat membranous fronds of *Chondrus crispus* occur in a variety of colours even with tinges of green or yellow. Dichotomously-branched and sub-divided, up to 15 cm (6 in) long with a short stalk and a disc holdfast; lateral fronds range in width from 1.25 to 2.5 cm ($\frac{1}{2}$–1 in). Unlike *Gigartina stellata*, carragheen moss, the fronds are never in-rolled. It grows in a wide variety of places except in mud, and on occasions may even form a carpet growth. Occurs in rock-pools, on stones, large shells and so on. Widely distributed.

Corallina officinalis (Coralline or Coral Weed)

Family Rhodophyceae **Fig. 5(41)** Very common/ common

Branches and branchlets much more numerous than in Fig. 4(25), and exactly opposite. On close inspection the main stem and branches, at least, are seen to be a series of very tiny oval flattened bead-

Carragheen (*Chondrus crispus*).

like sections. Grows in stiff crusty tufts of up to 15 cm (6 in) tall; normally plum-coloured, but when subjected to exposure this is reduced to yellow or even white depending on the degree of sunshine and especially on upper middle shore. Prefers shady places and at times can produce a prolific growth carpeting rocky areas. Often fringes pools about 7 or 8 cm ($2\frac{3}{4}$–3 in) below water surface. Very widespread; also occurs on lower shore.

Cryptopleura ramosa

Family Rhodophyceae **Fig. 4(21)** Very common/ common

Frond thin and membranous, up to 20 cm (8 in) high and of a similar width; much divided, usually dichotomously but often very irregular; stem short or absent. Purplish-red to brownish-red, margin wavy; only lower portion of frond is veined. Grows on rocks and stones, also on species of *Laminaria* weeds. Due to water action the membranous frond may become rolled-up and indistinguishable. Widely distributed and common, particularly in south. Also occurs on lower shore.

Cystoclonium purpureum

Family Rhodophyceae **Fig. 4(21)** Very common/ common

Forms a soft bushy growth; main stem is irregularly branched with many branchlets and further divisions; it is considerably longer than the branches. Main stem and branches are cylindrical; branches and branchlets taper at either extremity. Dull purplish-red, but appears pinkish when underwater. Grows from a holdfast of fine-branched rootlets which is attached to rocks, especially in pools. In the summer months swellings occur along the branches as the fruiting-bodies begin to develop. Does not survive the winter months but dies back leaving little trace of its existence. Common during summer; also occurs on lower shore.

Dilsea carnosa

Family Rhodophyceae **Fig. 4(20)** Very common/ common

Produces thick, flat, dark red fronds in spring and summer. These are up to 38 cm (15 in) long and 12 cm (5 in) wide, rounded at apex, broadest about midway, and then gradually narrowing into a round stalk. Often damaged and misshapen by wave action. Could be mistaken for *Rhodymenia palmata*, dulse, which does not always have subsidiary pieces of frond growing from margin of main blade. Grows on rocks and stones. Occurs on both the middle and lower shore zones.

Dumontia incrassata

Family Rhodophyceae **Fig. 4(29)** Very common/ common

A slimy weed up to 30 cm (12 in), with irregular worm-like, blunt-ended branches and branchlets. These are shallowly constricted at points along length where variances in thickness occur; narrowest at end close to stem. Colour normally purplish-red when growing on rocks and stones in shady positions or in pools. Where exposed to sunlight, often pales to a yellowish-green. Entire plant has a gelatinous texture. Holdfast is a very small disc. Very common and widely distributed in spring and early summer.

Gastroclonium ovatum

Family Rhodophyceae **Fig. 4(31)** Very common/ common

Main stem is irregularly branched; no branchlets. Brownish-red, erect, and up to 20 cm (8 in) tall. Easily identified by the tiny clusters of smooth, hollow, elongated red processes which are produced around the upper part of stem and branches (these look very much like red grains of wheat). Emanates from a holdfast of thread-like rootlets attached to rocks and among seaweeds in rock-pools. A widely distributed species which also occurs on lower shore.

Gigartina stellata (Carragheen Moss)

Family Rhodophyceae **Fig. 4(23)** Very common/ common

This flat dichotomously-branched weed usually grows in tufts and often among other red seaweeds. The very dark brownish-red fronds are up to 20 cm (8 in) with inrolled margins. May be found growing alongside *Chondrus crispa*, carrageen, from which it is easily distinguished by virtue of its inrolled margins. Pustular fruiting-bodies may be dotted quite

numerously over the frond surface; these later develop into very tiny, almost leaf-like processes. It is
often covered with a spiny greyish encrustation produced by the polyzoan *Flustrella hispida*, or en-
crusted with white patches of the bryozoan *Mem-
branipora membranacea*. It is attached by a disc
holdfast to stones and rocks on both the middle and
lower shores. It is widely distributed.

Hildebrandia spp.

Family Rhodophyceae Very common/common

A hard, sharply-defined smooth encrustation which
occurs on damp shady rocks especially in rock-
pools. Crimson in colour and looks more like a
patch of enamel paint than a seaweed. Shiny when
moist, it becomes dull when dry. A white outer
ring of calcium carbonate develops as plant ages;
this spreads inwards towards centre as death approaches. It can be picked away from surface of
habitat by thumbnail. Similar, but greenish-black,
is the lichen *Verrucaria mucosa*.

Laurencia hybrida

Family Rhodophyceae **Fig. 4(24)** Very common/
common

A dark purple plant with strong main stem which
is alternately branched. Branches bear alternate
branchlets which are themselves further sub-divided.
Stem and main branches cylindrical; not thread-
like. Grows up to 15 cm (6 in) high and can be found
attached to rocks and limpets on the beds of shallow
pools. Looks superficially like a miniature Christmas tree. It is widely distributed around our coasts;
occurs on lower middle shore.

Laurencia obtusa

Family Rhodophyceae **Fig. 4(22)** Very common/
common

This normally pinkish-purple weed may vary in
colour to reddish-yellow. The stem and branches
are stiff and cylindrical, not hair-like. Branches and
branchlets are opposite and widely spread, becoming progressively shorter towards tips. Branchlets
are similarly sub-divided. Grows up to 15 cm (6 in)
tall and is attached to rocks and stones in pools. A
summer species found on lower middle and lower
shores. Wide distribution.

Laurencia pinnatifidia (Pepper Dulse)

Family Rhodophyceae **Fig. 5(42)** Very common/
common

When growing in pools and other shaded situations,
it is typically purplish-red with a tufted growth of up
to 30 cm (12 in). However, if in a more sunny site
on rocks, it may be yellowish-green or even paler,
and reduced to a carpet of short dense tufts. It
has a strong main stem; the branches, branchlets,
and further sub-divisions are flat and alternate.
Branches become progressively shorter towards tip
of main stem, the branchlets reduce in length similarly towards tips of branches. Plant very variable in
form, sometimes branches and branchlets are not
completely flat, or may even be inrolled. Maximum size reached in winter. Occurs during summer
months on lower middle and lower shores.

Lithophyllum incrustans

Family Rhodophyceae Very common/common

A rough-surfaced, pinkish-purple encrustation of
abrasive texture. Very thin, sometimes has a pattern
of more or less concentric rings. On occasions
patches of *Lithophyllum incrustans* cover extensive
areas of large rocks. In no way does it resemble a
seaweed in the accepted sense. Colour tends to fade
towards edge of patches when growing on rocks;
and also in patches growing at water's surface in
pools. Plant turns white when dead. Seems to be less
common in very still waters. Widely distributed;
occurs on both middle and lower shores.

Lomentaria clavellosa

Family Rhodophyceae **Fig. 4(22)** Very common/
common

The hollow main stem is cylindrical. Branches,
branchlets and sub-divisions are opposite; they
become progressively shorter in length upwards.
Terminal sub-divisions have well-defined points.
Branches and branchlets taper at either extremity.
The entire plant is very pink, rather limp, and feels
somewhat gelatinous when handled; it has a ten-
dency to lie quite flat in one plane. Occurs during
summer on rocks and other seaweeds in pools.
Widely distributed on both middle and lower
shores.

Membranoptera alata

Family Rhodophyceae **Fig. 5(43)** Very common/common

This blood-red to pinkish-crimson weed consists of narrow, flat, ribbon-like, irregularly-dichotomous branches which bear dichotomously-tipped branchlets. A midrib runs through the entire plant, but there are no lateral veins. Grows up to 20 cm (8 in) long on shaded rocks amongst other seaweeds, and in pools attached to the stalks of *Laminaria* species of weed. Widely distributed and occurs on both the middle and lower shore zones.

Plumaria elegans

Family Rhodophyccac **Fig. 5(44)** Vcry common/common

A beautiful and delicate dark crimson weed which grows up to 10 cm (4 in) tall. The plume-like branches and branchlets have fine sub-divisions which are not obvious and must be viewed with a hand lens. When out of water the plant hangs limply from rocks with no semblance of its true form; this only becomes apparent when floating in water. Grows on rocks especially in shaded situations. Do not confuse with *Ptilota plumosa* whose stem and main branches are flattened and much more prominent than those of *Plumaria elegans*. Occurs on both the middle and lower shore zones.

Polysiphonia fastigiata

Family Rhodophyceae **Fig. 4(28)** Very common/common

There are over twenty species of *Polysiphonia*, many of which may answer this description. All are small with thread-like stems, branches and branchlets. The branches and branchlets have a very irregular and entangled tufted growth. The plant varies from dark red to purplish-red and grows up to 15 cm (6 in). Usually grows on other seaweeds, for example *Ascophyllum nodosum*, knotted wrack; it may cover much of its host but is not a parasite.

Rhodochorton spp.

Family Rhodophyceae Very common/common

A mass of dull crimson hair-like threads mixed with sand, and producing a somewhat fluffy or velvety nap or carpet-pile of up to 1.25 cm ($\frac{1}{2}$ in) tall. Covers substantial areas on rocks with species of other seaweeds growing up through it; may also grow on seaweeds. Widely distributed and may be locally abundant.

Rhodymenia palmata (Dulse)

Family Rhodophyceae **Fig. 4(19)** Very common/common

A tough, reddish-brown, stalkless frond which in water appears to have purple tints. Up to 30 cm (15 in) long and grows from a disc holdfast attached to rocks or the stalks of *Laminaria* and *Fucus* species. Very variablc in both shapc and sizc; sometimes just a single blade, either broad or narrow; at other times palmate; often has small finger-like lobes growing along margin. May occur in small pieces or bunched together. This is the edible dulse. Widely distributed on both the middle and lower shore zones.

Gelidium corneum

Family Rhodophyceae **Fig. 4(45)** Fairly common

The stem and main branches of this very dark-red weed are comparatively thick, stiff, and well developed. It carries branchlets and numerous sub-divisions. *Gelidium corneum* grows up to 15 cm (6 in) tall and the main branches are about 7 cm ($2\frac{3}{4}$ in) long, roughly half the length of main stem. Grows on rocks and is often quite prolific around edges of shallow rock pools. Widely distributed and occurs on lower middle shore. Fig. 5(45a) shows *G. latifolium*.

Hypoglossum woodwardii

Family Rhodophyceae **Fig. 4(18)** Fairly common

This pinkish-red weed has a ribbon-like frond from whose midrib leaf-like branches grow outwardly at right-angles. Similarly, leaf-like branchlets are produced at right-angles from the midribs of the branches. There are no lateral veins. Grows up to 20 cm (8 in) tall and is attached to rocks or the stalks of *Laminaria* species of seaweed in pools. Widely distributed occurring on both the middle and lower shore zones. It is especially common in the south.

SEAWEEDS FOUND ON LOWER SHORE ZONE

Codium tomentosum
Described in middle shore section.

Chorda filum (Bootlace Weed, Sea Lace or Mermaid's Tresses)
Family Phaeophyceae **Fig. 4(32)** Very common/common

Also known as dead men's ropes. It is round in section and consists of a cord-like, unbranched, slippery frond which tapers to a pointed tip. It grows up to 8 m (26 ft) long, reaching its maximum length in summer. The fronds are only 3–6 mm wide and grow from a disc holdfast attached to small stones and rocks, particularly in shallow water and other sheltered sites. In summer months often plays host to the membranously-fronded weed *Desmotrichium undulatum*, and becomes quite slimy with a covering of the weed *Litosiphon pusillus*. At low tide large amounts may be found rolled together in size-able bundles. During the reproductive season in early summer the fronds are covered in spore cells. Widely distributed on both the lower shore and sub-littoral zones.

Fucus serratus (Saw Wrack)
Described in middle shore section.

Laminaria digitata (Sea Tangle or Oarweed)
Family Phaeophyceae **Fig. 3(9)** Very common/common

Produces large brown fronds of up to 3 m (10 ft) in length and 90 cm (35 in) wide. These are smooth,

Sea tangle (*Laminaria digitata*).

tough and rubbery with long finger-like divisions. The long *flexible* stalk is *oval* in section (not *round* as in *L. hyperborea*). It grows from an holdfast of intertwined rootlets; these are *not thick* but thin and spread out flat over rock substrate. On exposed coasts it is often densely carpeted over rocks at water's edge at low tide; is also present in the larger rock-pools. It is widely distributed as a species and occurs on both the lower shore and sub-littoral zones.

Laminaria hyperborea (Forest Kelp)
Family Phaeophyceae **Fig. 3(9)** Very common/common

Best differentiated from previous species *L. digitata* by the following features. Stalk is *stiff* not flexible; is *round* in section not oval. Intertwined rootlets of holdfast are *short* and *thick*, and do not spread out flatly over rock substrate. The two species occupy similar areas on both the lower shore and sub-littoral zones.

Laminaria saccharina (Sea Belt or Sugary Wrack)
Described in middle shore section.

Leathesia difformis
Described in middle shore section.

Litosiphon pusillus
Family Phaeophyceae **Fig. 3(7)** Very common/common

Grows in tufts of slimy, yellowish-brown, un-branched threads on the fronds of other weeds especially *Chorda filum*, bootlace weed. The small tuft-like fronds grow up to 10 cm (4 in) long. Although it is only present during summer months, it is widely distributed on both lower shore and sub-littoral zones.

Mesogloia vermiculata
Described in middle shore section.

Punctaria plantaginea
Described in middle shore section.

Saccorhiza polyschides (Furbelows)
Family Phaeophyceae **Fig. 3(9)** Very common/common

Easily differentiated from the two digitate species of *Laminaria* already described by the following features. Frond deeply, almost completely digitate, usually has about eight long strap-like fingers. Stalk flattened, twisted, and has broad wavy frills along lower part of margin. An extremely massive and fast-growing plant, reaching lengths of almost 5 m (16 ft) and up to 2 m (6 ft) wide when growing in deep water; but more commonly about 2 m (6 ft) long. Holdfast is a large, hollow, roundish and knobbly structure attached to rocks; it may be up to 30 cm (12 in) in diameter. An annual which grows in isolation; frond is removed by the action of stormy winter seas. Widely distributed on both the lower shore and sub-littoral zones.

Ectocarous confervoides
Described in middle shore section.

Himanthalia elongata (Thong Weed)
Described in middle shore section.

Ahnfeltia plicata
Described in middle shore section.

Ceramium acanthonotum
Described in middle shore section.

Chondra dasyphylla
Family Phaeophyceae Very common/common

Main stem tough and usually undivided. Branches alternate, more or less parallel, becoming progressively shorter towards top of main stem. The branches have many short alternate branchlets; those on the longer lower branches are themselves almost as long as the topmost branches. Branches and branchlets are blunt-ended, narrowest at point of emanation. This purplish-red weed grows in pools, especially those with a muddy or sandy bed. Only occurs in summer; widely distributed.

Corallina officinalis (**Coralline or Coral Weed**)
Described in middle shore section.

Cryptopleura ramosa
Described in middle shore section.

Cystoclonium purpureum
Described in middle shore section.

Delesseria sanguinea
Family Phaeophyceae **Fig. 5(46)** Very common/
common
This many-branched, short-stemmed weed is deep
reddish-pink and grows up to 25 cm (10 in) tall. It
produces large elongated leaf-like fronds; each is
stalked and has a wavy but undivided margin. The
main-rib is prominent as are the lateral veins. Looks
much more like a 'tree-leaf' than a seaweed frond.
Grows more usually in isolation deep down in large
shaded pools; on rocks and occasionally on species
of *Laminaria*. Has a wide distribution and is in its
prime during summer months.

Dilsea carnosa
Described in middle shore section.

Gastroclonium ovatum
Described in middle shore section.

Gigartina stellata (**Carragheen Moss**)
Described in middle shore section.

Griffithsia flosculosa
Family Phaeophyceae **Fig. 4(26)** Very common/
common
A rather delicate weed with numerous dichotomously-branched threads. When under water it has
an erect tufted habit, the branched threads growing
closely together. Tiny stalked pustules are scattered
about the threads; these are the reproductive parts.
Grows in rock-pools, especially those on the more
exposed areas of our coasts. A widely-distributed
species.

Heterosiphonia plumosa
Family Phaeophyceae **Fig. 4(24)** Very common/
common
Main stem is tough, hairy and up to 30 cm long. It is
slightly angled away from the base of each alternate
branch; this produces a mildly zig-zagged effect.
Branches are bare at base before giving rise to alternate branchlets; these reduce in length towards tip
of branches. The branchlets too produce further
extra fine sub-divisions. This arrangement of
branches, branchlets and finer divisions makes for a
feather like effect, and the entire plant looks like a
deep crimson fern. Grows on rocks and stones, or
on other seaweeds in deep water *below* lower shore
zone, but large fragments are often to be seen
washed up on lower shore. Widely distributed.

Laurencia obtusa
Described in middle shore section.

Laurencia pinnatifida (**Pepper Dulse**)
Described in middle shore section.

Lithophyllum incrustans
Described in middle shore section.

Lithothamnion spp.
Family Phaeophyceae Very common/common
A very thin pinkish-purple encrustation which
grows on rocks in pools and sometimes on other
seaweeds. Only when the encrustations are lobed in
the manner of liverwort, or have spiky prominences,
are they readily distinguishable from *Lithophyllum*
species. At other times they can only be determined
on microscopic characters.

Lomentaria articulata
Family Phaeophyceae **Fig. 4(25)** Very common/
common
An irregularly-branched pinkish-red to crimson

weed up to 20 cm (8 in) long. The stem, branches and sub-divisions are constricted at regular intervals giving an appearance of small, shiny, closely-strung, longish, ovoid beads. Grows in pools on rocks, shells and other seaweeds. One of our commonest red weeds with a wide distribution.

Lomentaria clavellosa
Described in middle shore section.

Membranoptera alata
Described in middle shore section.

Odonthala dentata
Family Phaeophyceae Very common/common

Stem, banches and branchlets are firm and solid, but flat and narrowly ribbon-like. The branches are alternate as are the stumpy, terminally-toothed branchlets; tiny fruiting bodies occur between the teeth in winter. Bears a vague resemblance to Fig. 4(18), but branches and branchlets emanate from edges not from midribs. This red weed grows up to 25 cm (10 in) tall on rocks and shells in pools. It is often cast up on beach by incoming waves. Common in northern England and Scotland.

Phycodrys rubens (Seaweed Oak)
Family Phaeophyceae **Fig. 3(16)** Very common/common

The strong stiff stem of this brownish-red weed is many-branched and bears large stalked leaves. These leaves usually have lobed margins and look very like oak leaves; at other times the indentations are quite pointed and thus bear a superficial resemblance to holly leaves. All the leaves have prominent midribs and lateral veins. Grows up to 25 cm (10 in) tall, occurs in shady pools and more usually on *Laminaria* stalks. A widely-distributed species which is regularly cast up on the beach.

Phyllophora membranifolia
Family Phaeophyceae **Fig. 5(47)** Very common/common

It has a long round continuous stalk, from one side of which emanate slender branches. These branches terminate abruptly into several flat, somewhat triangular-shaped, membranous dichotomies which together produce a fan-like effect. This purplish-brown to purplish-red weed has a tiny disc holdfast and grows up to 25 cm (10 in) tall. Occurs in pools and is widely distributed.

Plocamium coccineum
Family Phaeophyceae **Fig. 5(48)** Very common/common

Stem and main branches well developed and somewhat wavy. Branches are irregular but main ones tend to be alternate. Branchlets are short and occur consecutively, and approximately in sequences of four, first on one side of the branch and then on the other. The further sub-divisions are really quite minute and these occur only on the outside of the branchlets. Very often lower parts of branches are without branchlets thus giving the plant a bare spindly appearance. This red weed grows up to 20 cm (8 in) tall and occurs on rocks in pools. It is very often cast up higher on shore. It has a wide distribution.

Plumaria elegans
Described in middle shore section.

Ptilota plumosa
Family Phaeophyceae **Fig. 5(49)** Very common/common

Main stem and branches are flattened, irregularly-branched, and not soft as in *Plumaria elegans*. Entire length of main stem, branches and branchlets is plumose; the fine sub-divisions of branchlets are set so closely together that their appearance is that of a rounded feather. A dark red to crimson weed which grows up to 20 cm (8 in) and is attached to other seaweeds, typically on the stalks of *Laminaria* species. Occurs principally on our northern coasts where it is common. Elsewhere uncommon.

Rhodymenia palmata (Dulse)
Described in middle shore section.

Halopitys incurvus
Family Phaeophyceae **Fig. 5(50)** Fairly common
This red weed has a tough main stem and is irregularly-branched. The branches produce numerous upward-curving branchlets on one side only, presenting an almost comb-like look. It forms thick tufts of up to 25 cm (10 in) tall. Rigid when dry. Grows on rocks and stones; has an holdfast of branched rootlets. Only occurs on south coasts where it is locally common.

Hypoglossum woodwardii
Described in middle shore section.

SEAWEEDS FOUND IN SUB-LITTORAL ZONE

Laminaria digitata **(Sea Tangle or Oarweed)**
Described in lower shore section.

Laminaria hyperborea
Described in lower shore section.

Laminaria saccharina **(Sea Belt or Sugary Wrack)**
Described in lower shore section.

Litosiphon pusillus
Described in lower shore section.

Saccorhiza polyschides
Described in lower shore section.

Seals

Grey Seal (*Halichoerus grypus*)

Three distinct populations of grey seal are recognised; these are the Baltic, the eastern Atlantic, and the western Atlantic. It is the eastern Atlantic population which is present in British waters. They come ashore to breed in the autumn, but there are a few exceptions; for example, along the coasts of Cornwall and Pembrokeshire a few pups are born in the spring. Eastern Atlantic grey seals often form large 'rookeries' in which the cow to bull ratio varies between 5:1 and 10:1.

The other two populations tend to breed on ice during late winter or early spring.

Over 60 per cent of the world's population live around our coasts; this represents a total of 80–85,000. The female is darkly and boldly blotched on a palish background, she has a slender snout and grows to about 2.1 m (7 ft). The male is of a more uniform darkish-grey, less prominently blotched,

Common seal (*Phoca vitulina*) on Shetland.

has a heavy 'Roman-nose' type of snout, and grows up to 2.4 (8 ft) in length. The grey seal is therefore Britain's largest carnivorous mammal. When fully adult, bulls weigh about 150 kg (3 cwt) or even up to 200 kg (4 cwt); cows about 100 kg (2 cwt). Females live for up to 30 years and males up to 20 years.

Although there is a large colony on the Farne Islands, their principal breeding colonies are more or less confined to areas along the west coast and the Scottish Islands; especially the Hebrides, the Orkneys and Shetlands; to a far lesser extent elsewhere, including islands off the Pembrokeshire coast and the Scilly Islands.

Common Seal (*Phoca vitulina*)

The name 'common seal' might suggest that it outnumbers the grey seal around our shores, but this would only be true in areas on the Norfolk coast and the Wash. In fact it is far less numerous in other coastal waters, with the total British population put at about 20,000.

It is circumpolar in the northern hemisphere, but five sub-species are recognised; the nominate inhabits European waters.

The common seal has a rounded head and a comparatively short snout with V-shaped nostrils; compare with the 'Roman-type' nose of the Grey Seal. They are medium sandy-coloured above, paler beneath, and blotched with darker brown; much more slender than the grey. The adult male is between 1.5 and 2 m (5–6½ ft) long, the female 1.3 to 1.5 m (4¼–5 ft); has a maximum weight of 114 kg (2¼ cwt). Common seals breed along the southwest coast of Scotland, the Hebrides, Orkneys and Shetlands, one or two sites on the east coast of Scotland, also north-east coast of Ireland; but principally around the Wash and Norfolk coast. They produce their young during June and July, coming

Female grey seal (*Halichoerus grypus*) on Shetland.

to rest on sand-banks, rocks, or ashore on the mainland, where they give birth to a single pup. The pups are well able to swim and dive from birth and are suckled both in and out of the water.

Types of Rock

Rock type	Brief description	Locations of good examples
BASALT	Dark reddish-brown crystalline rock frequently occurring as sheer vertical cliffs, possibly giving rise to curious shapes.	Giant's Causeway in N. Ireland, the isles of Skye, Muck and Eigg, the Farne Islands.
CHALK	A pure, soft limestone, opaque white in colour. Normally found as vertical or almost vertical cliffs.	Mainly eastern and southern England, especially Kent, Dorset, Sussex, Lincolnshire and Yorkshire.
GRANITE	A coarse-grained igneous rock which can be very variable in colour, pinkish-grey, whitish-grey or even bluish-grey. Where the sea pounds against the rock, many caves and coves are formed.	Cornish coast, Scilly Isles, Peterhead, Isle of Rhum.
LIMESTONE	Quite a soft, sedimentary rock, usually pale-coloured. Readily eroded by the sea's action and normally found as sheer sloping cliffs with stacks or reefs close by.	Great Orme's Head, Tenby Peninsula, Durham coast, Durness in Scotland.
SAND AND CLAY	Vary in colour from pale grey to blue, very unstable and constantly collapsing because of either rain or dry weather which causes them to dry out and crack.	Common along the Norfolk and Suffolk coastline.
SANDSTONE	A hard coarse rock made up of compacted quartz, variable in colour from a greenish-grey to the more common brownish-red, and often has a striped appearance. Arches, gullies and stacks are created, often with some beautiful effects.	Caithness, Pembrokeshire, including the offshore island of Skokholm.

Initial numbers refer to locations on Map 1 on page 76.

1. LUNDY ISLAND (BRISTOL CHANNEL)

Owned by the National Trust, this 4.8-km (3-mile) long island in the Bristol Channel is some 18 km (11 miles) offshore. Cliffs 120 m (400 ft) high encircle the island, and are the homes of numerous seabirds during spring and summer.

Although the name Lundy is Norse for 'puffin island', few of these birds breed on the island these days. There used to be a small gannetry where the lighthouse now stands at the northern end, but the birds left the site when the lighthouse was built.

Apart from the numerous seabirds, the island also has resident flocks of wild goats and Soay sheep. Below the high cliffs, grey seals can be seen on occasions.

Short visits and longer stays are possible and enquiries should be directed to the island administrator; telephone Woolacombe 870870.

2. GODREVY ISLAND (CORNWALL)

Prior to the erection of the lighthouse in 1859, this part of the Cornish coast was a notorious blackspot for shipping and claimed many lives. The lighthouse has been unmanned since the 1930s and was one of the first automatic stations to be put into service by Trinity House.

Much of the mainland in the area is owned by the National Trust and a good road leads to Godrevy Point where there is ample parking. St Ives Bay is a popular surfing area, but strong currents make it a dangerous location at times. A superb walk along the clifftops to Navax Point gives fine views across to Godrevy Island, on which hundreds of seabirds nest in summer. Below the point, grey seals breed in the caves, inside which choughs once nested until they were driven away from Cornwall, probably by jackdaws. The occasional fulmar and kittiwake use the rocky ledges both sides of Navax Point.

In late spring, sea thrift carpets the cliff-tops and the slopes leading down to the flatter rocks between the headland and the island. Care must be taken,

Godrevy Island and lighthouse from the mainland, showing sea thrift (*Armeria maritima*) in late May.

Map 1 Sea-cliffs, rocky islands and rocky shores. So much of the
north Scottish coast (shaded areas) is rocky cliffs and of outstanding
ornithological interest that it is very difficult to single out specific areas.

however, as the cliff-tops are not fenced off and the numerous tussocks can easily trip the unwary.

The point can be reached by turning left off the B3301 road through the small village of Gwithian, near the public house and just after passing over Red River, so named because it is stained red by waste from old tin mines.

3. LOWER PREDANNACK CLIFF AND MULLION ISLAND (CORNWALL)

Leased from the National Trust, this 16.5-hectare (41-acre) reserve is managed by the Cornwall Naturalists' Trust. The cliff-top flora is quite rich and breeding seabirds include kittiwake, guillemot, razorbill and shag, especially on Mullion Island.

The walk between Mullion Cove and Predannack Head is superb, and the view north along Mount Bay breathtaking on a clear day. Access to the reserve is unrestricted, apart from Mullion Island which is closed to the public. Parking is available in the old quarry beside the B3296 near Porth Mellin.

West of the Lizard in Cornwall, in the region of Kynance Cove.

4. KYNANCE CLIFF AND LOWER PREDANNACK DOWNS (CORNWALL)

The coastal scenery of this part of the Lizard Peninsula is breathtaking to say the least and consequently the area is very popular with tourists throughout the year, but especially during summer.

Parking is available just above Kynance Cove which can be reached by a narrow road from the A3083 main road to Lizard Point. Otherwise, the 86-hectare (215-acre) reserve is not too much of a walk along the coast path from Lizard Head.

The reserve flora is very interesting, with rarities such as the Cornish heath growing on the drier heathland. The car park is at the edge of the Kynance section of the reserve and paths lead northwards to Kynance stream which is the southern boundary of the Lower Predannack section.

The reserve is leased by the Cornwall Naturalists' Trust from the National Trust.

5. BLACK VEN (DORSET)

The precipitous sea-cliffs of Black Ven, now part of a 12-hectare (30-acre) nature reserve, are composed of grey lias clays containing a rich fossil fauna. It was here in 1811 that a 12-year-old girl named Mary Anning found the first complete fossil of an ichthyosaurus, a prehistoric reptile similar to a giant porpoise.

Frequent landslips make the reserve very dangerous in places and visitors should avoid the muddy slopes, especially in wet weather. Car parking is available in Charmouth. The reserve is reached by walking westwards along the beach.

The reserve is leased from the National Trust by the Dorset Trust for Nature Conservation to whom further enquiries should be made. The Trust headquarters are at 39 Christchurch Road, Bournemouth.

6. WHITENOTHE AND BURNING CLIFF (DORSET)

Overlooking Weymouth Bay, and forming the crescent of Ringstead Bay, this 46-hectare (115-acre) reserve is leased to the Dorset Trust for Nature Conservation by the National Trust. In all, it is composed of some 2.4 km (1½ miles) of cliff and foreshore along the coast from such famous landmarks as Durdle Door and Lulworth Cove. The white chalk cliffs of this stretch of the Dorset Coast are truly magnificent, and the 150-m (500-ft) cliffs on this reserve are no exception.

Breeding birds include herring gulls and jackdaws on the cliffs and the plant life consists of typical maritime species. Butterflies seen on the reserve include migrants such as the painted lady and clouded yellow.

Car parking is available at several places, including Ringstead, Holworth and South Down farm. Each one requires a walk of varying length to the reserve. Walking, away from the footpaths, can be dangerous.

7. THE NAZE (ESSEX)

Managed by the Essex Naturalists' Trust by agreement with Tendring District Council, this 8-hectare (20-acre) reserve is situated at the northern end of the large public open space on the cliff-top to the north of Walton-on-the-Naze. Just beyond the site is the Naze itself – a large area of shingle and saltings.

The 21-m (70-foot) high cliffs are interesting geologically, containing many fossil shells. The reserve is mainly grassland, with some thickets and pools which attract a variety of birds, including several migrants. The flora of the area is also quite interesting.

8. BEMPTON CLIFFS (HUMBERSIDE) RSPB RESERVE

A 6.5-km (4-mile) stretch of high chalk cliffs is approached via the B1229 Flamborough/Filey road and is signposted from Bempton Village.

The site provides nesting places for a large number of seabirds including a colony of gannets. Summer visitors include ring ouzel, merlin and wheatear. The pyramidical orchid is just one of over 200 recorded species of flower.

Access is obtained by a public footpath from the car park at the end of the cliff road. Observation points are provided and boat trips from Bridlington give splendid views of the birds nesting on the cliffs. The reserve is at its best from May to mid-July. Further information is available from the summer warden at Bempton Post Office, Bridlington, Humberside.

9. HAYBURN WYKE (YORKSHIRE)

This 13.5-hectare (34-acre) reserve managed by the Yorkshire Naturalists' Trust is leased from the National Trust. A great deal of the cliffs on the reserve is covered with dense scrub containing ash, oak, sycamore, hazel and holly. The flora includes woodruff, yellow pimpernel and hart's-tongue fern. Mosses and liverworts abound.

The reserve lies halfway between Scarborough and Robin Hood's Bay and parking facilities are available at the Hayburn Wyke Hotel by agreement with the owner.

10. BLACKHALL ROCKS (DURHAM)

The 12-hectare (30-acre) site is a good example of reef limestone of Upper Permian age in a sea cliff which is about 18 m (60 ft) high. Plant life is most interesting, with the rare sea spleenwort growing in rocky crevices and butterwort and round-leaved wintergreen in the damp hollows on the steeply sloping cliff-tops.

Access is via the A1086 Hartlepool to Peterlee road by turning eastwards at the crossroads near the southern end of Blackhall Colliery village, and forking south along a dirt track to the reserve. Park in the open space at the end of the track. The cliffs are dangerous in places and great care must be taken.

11. THE FARNE ISLANDS (NORTHUMBRIA)

Situated some 3 km (2 miles) off the Northumbrian coast, this group of rocky islands is of outstanding

Birdwatching on Bempton Cliffs.

ornithological interest. Thousands of visitors flock to them each year between April and September using the boat services from Seahouses. There is restricted access during the peak breeding months of May and June, but even then it is still possible to enjoy the delights of Staple Island, with its famous Pinnacles covered with guillemots.

In all, there are about 30 islands, with the furthest being 8 km (5 miles) off the coast. They contain a rich and varied wildlife, in particular the large colonies of seabirds, including kittiwake, fulmar, guillemot, razorbill, shag, cormorant, puffin, oyster-catcher, the various gulls and four species of tern – common, arctic, roseate and sandwich.

Grey seals are visible on the smaller rocky outlets

Farne Islands. Seabird colonies on cliffs white with accumulated droppings.

besides the larger islands and on their return to Seahouses visitors will not fail to notice the large population of eider duck using the coastline.

12. ST ABB'S HEAD (NORTHUMBRIA)

The wildlife reserve at St Abb's Head is managed jointly by the Scottish Wildlife Trust and the owners – the National Trust for Scotland. Over 76 hectares (190 acres) of sea-cliffs (composed of rocks from volcanic lavas over 400 million years ago) and coastal habitats form this reserve. It is located on the east coast of Scotland near the English border and can be approached from the A1107 Eyemouth road or from the B6438 road off the A1.

More than 50,000 seabirds nest here and include colonies of guillemot, razorbill, kittiwake, shag, fulmar and a few puffins. They are best viewed between April and late July. Common water birds can be found on the freshwater loch (Mire Loch) inland from the cliffs, and the headland supports stonechats, wheatears and meadow pipits.

Plant species include heather, tormentil, heath bedstraw, rock rose, lady's bedstraw, purple milk

St Abb's Head.

Bass Rock.

vetch and spring sandwort. Maritime plants found on the sea-cliffs include sea thrift, sea campion, rose-root and Scots lovage.

Butterflies recorded on the reserve include common blue, northern brown argus, grayling and small copper. In summer, the six-spot burnet moth is common.

Cars can be parked at Northfield Farm to the south of the reserve, almost 1 km ($\frac{1}{2}$ mile) west of St Abb's village. Visitors should then follow the coastal footpath to the lighthouse. A further car park is provided at the lighthouse itself. The reserve is open to the public throughout the year and further information can be obtained from the Ranger, Ranger's Cottage, Northfield, St Abb's (Tel: Coldingham 443).

13. BASS ROCK (LOTHIAN)

Bass Rock is best known for its large gannetry which dominates the island during spring and summer, giving it an impressive pale colouration more typical of the chalk cliffs of the south coast of England. Other breeding birds include razorbills, puffins, fulmars and cormorants.

Boat trips to the island are available from the harbour at North Berwick.

14. ISLE OF MAY (FIFE)

Situated about 8 km (5 miles) offshore, the towering cliffs of the Isle of May are home to numerous seabirds during spring and summer. It is also an excellent place to observe migration during the spring and autumn passage.

Breeding birds include up to 12,000 pairs of puffin, 15,000 pairs of guillemot and significant numbers of razorbills, kittiwakes, shags and both herring and lesser black-backed gulls. In late spring the flora is spectacular with abundant growths of thrift and campion all over the island.

Boat trips to the island are available from Crail and Anstruther.

15. SEATON CLIFFS (TAYSIDE)

Situated just to the north of Arbroath, this 10.5-

A school party arriving on the Isle of May.

hectare (26-acre) reserve consists of Old Red Sandstone cliffs along a mile of shore, including a strip of vegetation along the top. The coastal scenery is spectacular here and a Trust booklet is available outlining the Arbroath Cliffs Nature Trail. This is sold in local shops.

Cars should be parked at the north end of Arbroath promenade.

16. FOWLSHEUGH (GRAMPIAN) RSPB RESERVE

Three km (2 miles) of grass-topped cliffs provide homes for huge colonies of seabirds, including guillemots (30,000 pairs), razorbills and fulmars. Seals can be seen offshore.

The cliffs, an RSPB reserve, are reached along the cliff-top path northwards from the car park at Crawton, which is signposted from the A92 road 5 km (3 miles) to the south of Stonehaven.

There is no visiting charge to the reserve and further information can be obtained from the RSPB Scottish Office, 17 Regent Terrace, Edinburgh EH7 5BN (Tel: 031 556 5624).

17. COPINSAY (ORKNEY) RSPB RESERVE

The island of Copinsay is located just off the east coast of mainland Orkney. It covers some 150 hectares (375 acres) and is composed of Old Red Sandsone, with sheer cliffs some 1.5 km (1 mile) long. The cliffs provide homes for many seabirds, including kittiwake, razorbill and guillemot. However, outside the breeding season (late April to mid-July) the island attracts very few birds except in adverse weather conditions. Plant species include oyster plant, sea aster and sea pearlwort.

Copinsay is reached by boat from Skaill (S. Foubister), for which there is a charge. The boatman can be contacted by telephoning Deerness 252. Further details about the island can be obtained from the RSPB Orkney officer Eric Meek, Cromlech, Stenness, Stromness, Orkney KW16 3JX (Tel: 0856 850176).

18. NORTH HILL, PAPA WESTRAY (ORKNEY) RSPB RESERVE

The RSPB reserve, covering just over 200 hectares (500 acres), is located at the north end of the island's main road. It has a rocky coastline with some low

Fowlsheugh Cliffs, near Crawton.

cliffs and is backed by maritime sedge heath.

Bird life includes arctic tern, arctic and great skua, corncrake and puffin. Plant species include thrift, sea plantain, crowberry, Scottish primrose, Alpine meadow rue, Alpine bistort, slender bed-straw and frog orchid.

There is no visiting charge, but visitors are asked to contact the warden on arrival at the cottage at Gowrie, 100 m east of the reserve entrance. Best views of the breeding seabirds are from mid-May to July, and the summer warden will give tours of the reserve if given prior notice. Further information is available from the summer warden, c/o Gowrie, Papa Westray, Orkney KW17 2BU, or from the RSPB Orkney Officer.

19. NOUP CLIFFS (ORKNEY) RSPB RESERVE

Some 2.5 km (1½ miles) of high cliffs with horizontal layers of Old Red Sandstone provide the site for the second largest colony of nesting seabirds in Britain. The cliffs which are backed by heathland are located on the west coast of the island of Westray and are approached from Pierrowall by taking the minor road west to Noup Farm and then the track north-west to the lighthouse.

Breeding colonies of guillemot, kittiwake, razor-bill, puffin and fulmar occur from May to mid-July. Grey seal, dolphin and porpoise are often sighted.

The reserve is open at all times and there is no visiting charge. Further information is available from the RSPB Orkney Officer, Eric Meek, Crom-lech, Stenness, Stromness, Orkney KW16 3JX (Tel: 0856 850176).

20. MARWICK HEAD (ORKNEY) RSPB RESERVE

One-and-a-half km (1 mile) of steep cliffs with hori-zontal beds of red sandstone form a breeding ground for seabirds including 35,000 guillemots,

10,000 kittiwakes, fulmars and razorbills. The birds provide a splendid scene from late April to mid-July, but the cliffs are deserted at other times. Plant species include sea thrift, sea campion and spring squill – all three flower together, making late spring the best time to visit the reserve for sheer beauty. Otters, and grey and common seals, are regularly seen.

This RSPB reserve lies to the west of Dounby on a minor road from the B9056. It is reached by leaving cars parked at the road end and walking the path northwards from Marwick Bay. A car park at Grid Reference 232252 gives access to the northern part of the reserve. There is no visiting charge and fur-ther details can be gained from the RSPB Orkney Officer, Eric Meek, Cromlech, Stenness, Stromness, Orkney KW16 3JX (Tel: 0856 850176).

21. NOSS (SHETLANDS)

Located to the east of Bressay in the Shetlands, this small island forms a National Nature Reserve of considerable importance for seabirds. Horizontal ledges on the high sandstone cliffs on the south and east shores provide an ideal breeding habitat for one of Europe's largest seabird colonies (an estimated 70–80,000 birds). Bird species include great and arc-tic skuas on the moorland at the centre of the island; eider ducks on the shore; kittiwakes, gannets, puf-fins and guillemots on the cliffs.

Plants include sea and red campion, spring squill, thrift, roseroot and mayweed. The island is un-inhabited except for the house used in summer by wardens. A suggested walk around the island takes two to three hours and includes the 180-m (600-ft) Noup of Noss cliffs which, on a fine day, give mag-nificent views over the whole of the Shetlands. Visi-tors are asked to keep away from the centre of the island to minimise disturbance to wildlife.

Further information may be obtained from the Nature Conservancy Council, 12 Hope Terrace, Edinburgh EH9 2AS.

22. FETLAR (SHETLANDS) RSPB RESERVE

Fetlar lies to the east of Yell and to the south of Unst, and a public car ferry runs from both islands. About 680 hectares (1700 acres) of the island forms the RSPB reserve and a large proportion of this acreage consists of moorland. The coastline is varied consisting of cliffs and boulder and sandy beaches.

Storm petrel, arctic and great skua, snowy owl, auk, manx shearwater and merlin are just a few of the many bird species to be seen. Otter, common and grey seal are also seen regularly. Plants include dwarf willow, bog asphodel and spring squill, but over 250 species have been recorded.

Fetlar is inhabited and hence accessible throughout the year. The bird reserve, however, is closed from mid-May to late July, except to see the snowy owl by arrangement with, and escorted by, the warden. There is no visiting charge but photographers are asked to consult the warden before erecting hides. Advice on camping sites should be sought from the warden on arrival. The warden can be contacted at Bealance, Fetlar ZE2 9DJ (Tel: Fetlar 246). The island shop closes on Sundays and Mondays.

23. BALRANALD (WESTERN ISLES) RSPB RESERVE

This RSPB reserve is situated on the west coast of North Uist and access to it is via the A865 road from Bayhead. The reserve contains a range of habitats, including rocky coastline, sandy beaches and dunes, machair, marshland and lochs.

Nesting waders are typical of the machair, on which smaller birds like twite, corn bunting and wheatear also breed. Eider, fulmar, black guillemot and shelduck nest along the coast, where grey seals are present throughout most of the year, especially on the small offshore island of Causamul.

The reserve is open daily from April to September, but visitors are asked to go to the Visitor's Cottage at Goular on arrival. The building is also available for shelter during bad weather.

24. HANDA (HIGHLANDS) RSPB RESERVE

The island of Handa lies to the north-west coast of Sutherland, about 1.5 km (1 mile) by sea from Tarbert and 5 km (3 miles) from Scourie. Handa is an RSPB reserve covering over 300 hectares (750 acres) and comprising sandy bays on the south and 120-m

Handa Island, capped by cloud, taken from a hillside south of Tarbert.

(400-ft) sandstone cliffs on the north side, with rough pasture in the centre of the island. The cliffs provide nesting sites for large numbers of kittiwake, guillemot, razorbill, puffin and fulmar. In all, over 150 bird species have been recorded. Grey seals, dolphins, otters and occasionally killer whales may be seen.

Plants found on the island include northern marsh orchid, spotted heath orchid, bog asphodel, limestone bugle, lovage and pale butterwort. Amongst the butterflies, small tortoiseshell, meadow brown, large heath and common blue frequently occur in numbers.

Between April and August, boat trips are available daily (except on Sundays) from Tarbert to the island. There is a small landing charge for non-members. Day visitors can use the stone-built shelter by the beach at Port an Eilein. Camping on the island is prohibited. Further details are available from the summer warden, Mrs A. Munro, Tarbert, Foindle, By Lairg, Sutherland IV27 4SS. The boatman can be contacted on Scourie 2156.

25. EIGG RESERVES – ISLE OF EIGG (HIGHLAND)

Some 1,500 hectares (3,750 acres) of the wilder parts of Eigg are designated nature reserves managed by the Scottish Wildlife Trust. There are three parts in all, each one offering a range of habitats including the ridge of An Sgurr, small lochs, willow scrub, cliffs of basalt rock and mixed woodland.

The resident population of Eigg is no more than about 80, but accommodation for visitors to the island is available in guest houses. Access to the island is by MacBraynes Ferry from Mallaig or by the regular service from Glen Uig. A special permit is needed to enter the reserves and this is available from the Estate Office, Isle of Eigg, where other information about accommodation can be obtained.

The bird and plant life on the island is rich and varied, with a nesting colony of Manx shearwaters. A leaflet and map are available from the Scottish Trust, 25 Johnston Terrace, Edinburgh.

26. MULL OF GALLOWAY (DUMFRIES AND GALLOWAY) RSPB RESERVE

This RSPB reserve of steep cliffs lies at the southernmost tip of the Rhins Peninsula and is approached by the A716 to Drunmore and then the B7041 road to the lighthouse. Rock ledges provide nesting sites for guillemot, razorbill, fulmar, great black-backed gull and herring gull. Gannets are often sighted fishing offshore.

Flowers growing on the cliff-tops include spring squill and purple milk vetch. There is no visiting charge and further information can be obtained from the RSPB Scottish Office, 17 Regent Terrace, Edinburgh EH7 5BN (Tel: 031 556 5624).

27. ST BEES HEAD (CUMBRIA) RSPB RESERVE

Twenty-two hectares (55 acres) of sheer sandstone cliffs form the St Bees Head RSPB reserve. The 90-m (300-ft) high cliffs lie south of Whitehaven and west of the B5345 road, and support over 5,000 breeding seabirds including puffin, razorbill, guillemot and kittiwake.

The reserve is reached from the footpath (with safe observation points) from St Bees beach car park. Cars are not permitted on the private road from Sandwith to the cliffs. The reserve is open all year and there is no visiting charge. The summer warden, c/o Leighton Moss Reserve, Myers Farm,

Silverdale, Carnforth, Lancs LA5 0SW, will supply further information.

28. SCARLETT VISITOR CENTRE (ISLE OF MAN)

The Scarlett Peninsula lies about 2.5 km (1½ miles) to the south-west of Castletown. A limestone cottage, derelict for many years, is now a visitor centre, housing the Manx Nature Conservation Trust's exhibition of information relating to the geology and natural history of the area. It is open on three afternoons per week between May and September. The Scarlett Nature Trail was the first permanent trail to be opened in the Isle of Man and it emphasises the geology of the area, which has a region of limestone, rare on an island dominated by slate.

One of the features of the area is the large Stack of Scarlett, composed of basalt rock – one of the chief volcanic rocks in the earth's crust. The stack's ledges are a favourite site for seabirds such as auks, shags and cormorants.

The cliff vegetation along most of the trail is typically maritime, and lichen flourishes almost everywhere.

The start of the trail can be reached by leaving Castletown Square along Queen Street and following the road round the bay. Car parking is plentiful just before Scarlett Quarry. The visitor centre is located on the other side of the quarry along the nature trail.

29. GREAT ORME (GWYNEDD)

Visitors to the resort town of Llandudno will immediately notice the formidable limestone promontory known as the Great Orme jutting out into the Irish Sea. The area is so popular in the summer months that a visit to the Little Orme across the bay may prove to be more rewarding for the bird-watcher.

Both sites offer a similar limestone habitat with sheer cliff-faces next to the sea. Care must be taken as the rocks are prone to become very slippery after a spell of wet weather.

The area is rich in flora including several rarities absent from other limestone regions in North Wales. Butterflies abound in summer, including a

The limestone cliffs of the Little Orme. Both the Great and Little Ormes are botanically rich as well as having seabird colonies.

diminutive sub-species of the grayling, which is found nowhere else in Britain.

On the sea-cliffs, breeding birds include shag, kittiwake, fulmar, razorbill, guillemot and herring gull. In recent years, the odd raven has nested, as well as a peregrine. Both sites are excellent for observing spring and autumn migrations.

A good road encircles the Great Orme, although stopping places for cars are few and far between. It is also a one-way system which can be a problem at times. The walk around the headland is not too strenuous and far more stimulating for the naturalist.

If solitude and tranquillity are what is required, then the Little Orme is well worth a visit. There is no recognised parking site, however, so that it is best to leave a car at the end of the promenade before the road begins to steepen. In June and July, the views both east and west are breathtaking and there is an abundance of flora and fauna to enjoy.

30. SOUTH STACK CLIFFS (GWYNEDD) RSPB RESERVE

The tall cliffs at South Stack are the summer home of large numbers of razorbills, guillemots and kittiwakes. Grey seals are regular visitors to the caves at the foot of the rocks and choughs always make an appearance somewhere along the cliff-tops.

Excellent viewing facilities are located inside the RSPB information centre, where the warden is in residence from April to September. The reserve is open daily and good parking facilities exist a few hundred metres from the site.

In fact, the walk along the cliff-tops in late spring and summer is a glorious one. Gorse and heather carpet the approaches and thrift and sea campion are prominent on the cliffs themselves.

To visit the reserve, follow the A5 to Holyhead where the road to the cliffs is well-signposted. The warden can be contacted at 'Plas Nico', South Stack, Holyhead, Anglesey.

31. LLANDDWYN ISLAND (GWYNEDD)

The rocky promontory of Llanddwyn Island (Ynys Llanddwyn) is part of the extensive National Nature Reserve dominating this part of the Anglesey coastline. The rocks are some 600,000,000 years old and are among the oldest rocks in Britain, belonging to the pre-Cambrian era of geological history. Both shags and cormorants breed on the small island just offshore from the main promontory. In late spring and early summer, spring squill is abundant on the grassy areas. At low tide, a good selection of seaweeds can be found and in the sheltered rock crevices delicate spleenwort ferns grow.

Access is either along the beach from the Forestry Commission car park in Newborough Forest or by foot along a footpath leading through the forest from the village of Newborough – a good 5 km (3 miles) long.

32. BARDSEY ISLAND (GWYNEDD)

Bardsey is the only sizeable offshore island in North Wales, unlike the south of the country where there are at least half a dozen islands rich in wildlife. Bardsey appears to be no more than a large rocky hump from the mainland at Aberdaron, but four-fifths of its 176 hectares (440 acres) are cultivated and low-lying. In fact, the illusion is caused by the position of the 170-m (550-ft) high mountain known as Mynydd Enlli which conceals the low-lying areas of the island.

Since 1953, when the Bardsey Bird and Field Observatory was established as a registered charity, thousands of visitors have been able to share in the delights of the island. It is the only accredited bird

observatory in Wales and, although great emphasis is placed on the ornithological value of the island, it is also of interest to botanists, entomologists, marine biologists and others; such people regularly visit Bardsey. There is a full-time warden on the island between March and November, assisted by several others, some of whom are volunteers. Some weeks are devoted entirely to courses, but most of the time the island is free for visitors to enjoy.

Bardsey is an excellent place to watch migration in spring and summer. By late March, the first chiffchaffs, goldcrests and wheatears are passing north, and later in April willow warblers, whitethroats and spotted flycatchers. The occasional rarity is recorded, especially during May, and the recent list includes golden oriole, scarlet rosefinch and red-rumped swallow.

Over 30 species nest on the island which is famous for its choughs. Five or six pairs currently breed there, although numbers do fluctuate. Other breeding seabirds include guillemots, razorbills, fulmars, gulls and the burrow-nesting manx shearwater.

On dark summer nights, the shearwaters' weird calls are clearly heard as they fly over the visitors' accommodation in Cristin farmhouse. Occasionally, visitors are allowed to accompany the warden and his assistants on nocturnal ringing expeditions.

August to October is the busiest period when common species appear in large numbers. August's

Bardsey Island, showing the lighthouse.

numerous warblers are followed in September by chats, flycatchers and goldcrests. October is vibrant, with continual passage of restless migrants; Bardsey is in fact one of the few places where night migration can actually be observed because the birds are attracted to the revolving beams of the lighthouse, often in their thousands.

Bardsey is a beautiful and interesting island, steeped in history. It offers the visitor solitude in a timeless atmosphere rich in wildlife. Further details about visits can be obtained from the Bookings Secretary, Mrs Helen Bond, 21A Gestridge Road, Kingsteignton, Newton Abbot, Devon. Boats to the island normally leave from Pwllheli.

33. PENDERI (DYFED)

Situated about 8 km (5 miles) south of Aberystwyth, this attractive 8-hectare (20-acre) reserve is composed of a stretch of cliff and steep sloping land, some of which is covered with oak trees. Flowers thrive on the grassy slopes and breeding seabirds include fulmar, herring gull, chough, cormorant and shag. Grey seals are regularly seen in the autumn.

Access to this West Wales Naturalists' Trust reserve is on foot from Pen-y-Graig farm (SN 553726) which is signposted from the A487 Cardigan to Aberystwyth road.

34. CARDIGAN ISLAND (DYFED)

As might be expected, Cardigan Island is situated some 6.5 km (4 miles) north-west of the town of Cardigan and is a nature reserve owned by the West Wales Naturalists' Trust.

The 16-hectare (40-acre) islet is grazed by a flock of Soay sheep. In summer, it is home for hundreds of seabirds, including shag, cormorant, oyster-catcher and the various gulls. Attempts are being made to attract manx shearwaters onto the island by moving fledglings from nearby Skomer. The West Wales Naturalists' Trust are hoping to attract puffins too, but using a different technique – setting up cardboard decoys on the cliffs early in the breeding season.

Landing is difficult even though the island is only some 400 m (1,300 ft) off the mainland. Hopeful

visitors should contact the Trust for further information (see Appendix 7).

35. DINAS HEAD (DYFED)

Often referred to as Dinas Island, the headland is in fact firmly joined to the mainland between Fishguard Bay and Newport Bay. The Pembrokeshire Coast Path goes right along the cliff-tops, which rise to nearly 150 m (500 ft) in some places. The interesting part about a walk around Dinas Head is that it starts and finishes at a pub, serving lunchtime meals in addition to the usual liquid refreshment. This probably explains why this part of the coastline is so popular with visitors to the area.

The scenery can only be described as magnificent. The range of birds seen is substantial at most times of the year. Raven, buzzard and peregrine breed in the area, as well as numerous seabirds such as fulmar, shag and razorbill. Choughs breed in the caves below the cliffs where seals can also be seen at certain times of the year.

As with most of the Pembrokeshire coastline, the range of wild flowers is impressive with spring squill and cowslips blending in nicely in late spring.

The site is approached by turning north off the main A487 (T) road from Fishguard to Cardigan in the village of Dinas, and aiming for Bryn-henllan. Parking is available near the cafe-cum-pub, a few hundred metres beyond the village besides a sandy cove known as Pwll Gwaelod.

36. STRUMBLE HEAD (DYFED)

This popular headland is approached by a network of narrow lanes leading off the A487 St David's to Fishguard road. Parking is available on the cliff-top opposite the lighthouse, which is joined to the mainland by a footbridge.

Excellent views exist to the east and west. In the breeding season, the cliffs are alive with noisy seabirds, notably auks, kittiwakes and herring gulls. Choughs are regularly seen, but only in small numbers. Grey seals regularly bask on rocks around the lighthouse itself.

The area is unrivalled walking country, with the famous Pembrokeshire Coast Path passing along the headland. Flowers abound in late spring and

Cliffs at Strumble Head.

The spectacular high cliffs and caves are in fact best seen from a boat. In fine weather, good views are available of nesting auks. The island boasts a large herd of grey seals with several hundred pups born each autumn.

Landing can be arranged and is well worth the effort. Resident birds include chough, peregrine, buzzard and raven. The absence of puffins is explained by the large rat population; rabbits too are abundant.

38. WEST HOOK CLIFFS (DYFED)

In an area of dramatic scenery, this 9-hectare (22-acre) reserve is leased by the West Wales Naturalists' Trust from the National Trust.

During spring and summer the cliff vegetation is outstanding, with spring squill growing amongst the more maritime species. Heath-spotted orchids grow in the damp area of the reserve and the rarer butterflies recorded include marsh fritillary and green hairstreak.

Parking is available at the National Trust car park at Martin's Haven and access to the cliffs is eastwards from Martin's Haven beach along the north side of the peninsula.

summer, attracting numerous species of butterflies. Both damselflies and dragonflies are found near the small stream which flows down to the sea just to the west of the car park.

In late May, the roadside along the approach to the site is covered with thrift and campion – a sight not to be missed.

37. RAMSEY ISLAND (DYFED)

Extending to some 260 hectares (650 acres) in all, this island is steeped in history with legends associated with it dating back to the sixth century.

Boat trips around the island run regularly in summer and are very popular with tourists. They begin at Porth Stinan which can be reached from St David's.

39. GRASSHOLM ISLAND (DYFED) RSPB RESERVE

The island is situated 16 km (10 miles) off the beautiful coast of Pembrokeshire, part of the new county of Dyfed. Like its neighbouring island Skomer, it consists of basaltic rock, but is much smaller in size, being only about 9 hectares (22 acres) in all. There is no soil to speak of on the island and no fresh water. Together with the fact that landing is very difficult except in the calmest weather, it is not surprising that the island is uninhabited.

Grassholm is owned and managed by the RSPB, who purchased the reserve in 1947. It is famous for its magnificent gannetry, with about 20,000 pairs breeding annually. Whilst this figure has grown from around 100 pairs in the 1920s, there has been an equivalent drastic decline in the number of puffins on the island, due mainly to the collapse of any burrows that existed; only a few pairs breed there presently.

Grassholm Island, showing the gannetry.

The island is an excellent place to see grey seals, with sometimes over a hundred basking on rocks during the summer months. Weather permitting, small bird migration observations can be very interesting in early spring and autumn.

Landings on the island are only permitted from mid-June onwards. This is to give as much protection as possible to the gannets at the crucial period of egg-laying, when the eggs are very vulnerable to predation by gulls. Sailings normally take place from Dale, and intending visitors should contact the West Wales Naturalists' Trust, 7 Market Street, Haverfordwest, Dyfed, for further details.

40. SKOMER ISLAND (DYFED)

Skomer is a National Nature Reserve leased by the West Wales Naturalists' Trust from the Nature Conservancy Council. The island lies just off the Marloes Peninsula and extends to some 288 hectares (720 acres) in all.

Skomer has the finest seabird colonies in southwest Britain, with fulmar, guillemot, kittiwake, puffin, oystercatcher and the three large species of gull breeding there. But its more renowned residents are the 100,000 or so pairs of manx shearwaters. Land birds include buzzard, chough, raven, wheatear, skylark, pheasant, little owl and short-eared owl.

Offshore, grey seals can normally be seen basking on reefs at low tide. Inland, bluebells and red campion are a magnificent sight in late May and early June. Maritime plants include sea campion and thrift. The island boasts the famous Skomer vole, an island race of the mainland bank vole. Rabbits are numerous too.

Skomer is open daily (except on Mondays), from April to late September. The official boat to land visitors on Skomer is operated by the Dale Sailing Company (Tel: Dale 349). Weather permitting, it leaves Martin's Haven at about 10 am daily and makes several journeys depending on demand. Visitors normally have about five hours ashore.

Limited accommodation is available on Skomer, but only for members of the West Wales Naturalists' Trust and assistant voluntary wardens. Full

information is available from the West Wales Naturalists' Trust, 7 Market Street, Haverfordwest, Dyfed.

Skokholm Island is popular with birdwatchers.

41. SKOKHOLM ISLAND (DYFED)

Lying just 3 km (2 miles) off the beautiful southwest Pembrokeshire coastline, Skokholm has an impressive range of red sandstone cliffs, together with several small bays and inlets. The cliff slopes support numerous seabirds and throughout summer are ablaze with bluebells, sea thrift and sea campion. The central area of the island is kept close-grazed by the large rabbit population. Many of the empty rabbit burrows are taken over by puffins and manx shearwaters during the summer, making several areas unsafe for walking without the possibility of numerous tunnels collapsing on the eggs or chicks inside. Some 35,000 pairs of shearwaters and 5,000 pairs each of puffins and storm petrels breed on the island, along with all three of the larger British gulls. Several species of land-birds nest there too. During peak migration periods, a large assortment of species may arrive, some quite rare.

Day-trips to the island are very infrequent, but week-long stays are available, and up to 15 visitors can be accommodated at any one time. Not all visitors are interested just in birds. Skokholm has an interesting flora, geology and shore-life. Also, courses are organised throughout spring and summer on subjects ranging from ornithology to art and photography.

The boat to the island leaves Dale at about 1 pm each Saturday, but bad weather can cause delays. Frequently, visitors are stranded for several days on the island and are asked to ensure that they are aware of this possibility. Further information is available from the West Wales Naturalists' Trust, 7 Market Street, Haverfordwest, Dyfed.

42. STACK ROCKS (DYFED)

Two spectacular sea-stacks, often referred to as the Elegug Stacks, stand out impressively from the coastline near Flimstone. The limestone pillars are breeding sites for numerous seabirds throughout spring and early summer, and are close enough to the mainland to offer visitors superb views of the comings and goings of the resident guillemots, razorbills, kittiwakes and fulmars.

A few hundred metres to the west, there is a classic example of a natural arch. The so-called 'Green Bridge of Wales' is probably the best such limestone feature in Wales.

Access to both sites is along a narrow road through a Ministry of Defence tank range. The roadway leads off the B4319 Castlemartin to Pembroke road, near the village of Warren. It is not always possible to visit the stacks due to the military activities. A large car park is available close to the edge of the cliffs, but children should be supervised because of the dangerous nature of the coastline.

Rocky shore showing mussel beds. Worm's Head area of South Gower coast.

43. WORM'S HEAD (WEST GLAMORGAN)

Situated at the southern end of beautiful Rhossili Bay, Worm's Head is a National Nature Reserve extending to some 480 hectares (1,200 acres) in all. It is a long rocky promontory linked to the mainland by a causeway which is only exposed at low tide, for four to five hours or so. The local coastguard station will provide details about safe crossing times.

The area is popular with botanists for the variety of limestone plants growing there, including several rarities. The steep cliffs at the far end provide nesting places for four different gulls, guillemot, razorbill and fulmar. The occasional puffin can also be seen, but not in the concentrations found on Skomer and Skokholm further to the west.

There is ample car parking available in the village of Rhossili and the walk down to the headland is a long, but pleasant one, with fine views across Rhossili Bay. Those who like exploring rock-pools will be delighted with what awaits them at the end of the path.

44. PORT EYNON (WEST GLAMORGAN)

Almost 2.5 km (1½ miles) of limestone cliffs along the South Gower coast make up the 100 hectares (250 acres) of nature reserves situated in this beautiful part of South Wales. The reserves have the most descriptive names, including Sedgers Bank, Longhole Cave Cliff and Deborah's Hole.

Over 200 plant species have been recorded, with hoary rock-rose and yellow whitlow grass among the rarest. Fulmars, kittiwakes and the occasional pair of ravens can be seen along the cliffs during spring and summer.

Parking is available some 300 metres north of the Port Eynon entrance to the reserves. Further information is available from the Glamorgan Naturalists' Trust, Tondu, Bridgend.

2

SHINGLE BEACHES, SANDY SHORES AND SAND DUNES

Introduction

Shingle deposits have led to the development of some unique and important shingle land forms around the coasts of Britain. In addition to the simple fringing beaches at the top of the shore, spits and bars have formed in several areas. A notable example of a spit is located at Orfordness in Suffolk, which extends down the coast for nearly 20 km (12 miles), directing the flow of the rivers Alde and Butley southwards before they enter the sea.

Chesil Beach, which extends for almost 30 km (18 miles) along the Dorset coast, is a classic example of a bar. This stretch of shingle completely encloses a lagoon, removing the influence of the tide and thus creating an interesting freshwater habitat for some unusual plant and animal communities.

Mudflats and saltmarshes are frequently formed in the sheltered waters behind shingle spits and islands. At Scolt Head and Blakeney Point on the north coast of Norfolk, a complex system has developed, where sand dunes cover certain areas in association with the saltmarshes and mudflats.

A suitable source of parent material is necessary for the development of shingle structures. Shingle is a mixture of varying sizes of stones and pebbles, ranging from 5 to 250 mm ($\frac{1}{5}$–10 in) in diameter, which were originally portions of rock from either the sea-bed or cliffs. Through centuries of erosion by violent wave-action, the pebbles have become rounded or semi-rounded. They may be unlike the local cliff or coastline and may drift parallel to the shore from elsewhere on the coast, often producing depositions on a considerable scale.

When a wave breaks, the strong uprush of water or 'swash' carries a mixture of sand and shingle up the beach. As a result, the larger coarser pebbles are left stranded at the landward margin of the beach whereas the finer material is dragged seawards by the weaker 'backwash'.

In this way, a build-up of large pebble embankments or storm beaches may occur on the more exposed coastlines. The 2.5-km ($1\frac{1}{2}$-mile) long pebble embankment at Newgale in Dyfed is a superb example of a storm beach. The well-rounded pebbles, worn smooth as a result of constant movement by the waves, mainly originate from nearby cliffs to the north-west and south. However, some are 'foreign', probably having been washed from glacial deposits flooring St Bride's Bay. At Dungeness in Kent, the large shingle deposits have built up as a series of successive storm beaches.

About a quarter of our coastline is bordered by shingle, but much of this is completely barren. A shingle beach is not stable enough to support the seaweed population typical of rocky shores. Often, they are steeply sloped and the pounding action of the waves creates enormous frictional forces between the pebbles preventing colonisation by both plants and animals. Drifting debris accounts for the only humus present on many shingle beaches, but this is often sufficient to support a minimum of flora and fauna. On the higher level areas, no longer disturbed by ordinary high tides, characteristic plants and animals do become established. As these die, the humus content increases and a certain element of stability arises, allowing typical shingle species to colonise vast expanses.

Vegetated shingle in Britain accounts for about 15–20,000 hectares (37,500–50,000 acres) and, apart from locations already mentioned, there are several other botanically important shingle sites around our coast, not so extensive perhaps as Dungeness and Orford Beach, but well worth a

visit. Many are protected by the Nature Conservancy Council as Sites of Special Scientific Interest.

Fortunately, shingle beaches are not as popular as sandy shores, but their limited recreational use undoubtedly helps to break up the stabilising vegetation. In several places, vehicles are given access to shingle sites, leading to the eventual weakening of the structure together with the loss in some areas of species of plants characteristic of shingle. The oyster plant is now common only in northern Britain, whilst the sea pea has vanished from more than half the sites where it used to grow. All of this has taken place relatively recently, perhaps in the last thirty years or so. Therefore, it is inevitable to link this decline to increased recreational activity and traffic volume during this period.

Another significant threat to the survival of any shingle structure is the extraction of gravel, normally from the sediment supply offshore. Such work is carried out at 'The Crumbles' near Eastbourne on the south coast. The creation of large open pits and the subsequent water abstraction may lower the water-table, possibly seriously affecting the shingle vegetation which is so dependent on an adequate water supply. However, at Dungeness, commercial gravel extraction in recent years has produced a completely new habitat feature. The pits created,

Vehicles allowed onto shingle ridges pose a threat to the flora and birdlife of this rare habitat.

together with their associated islands, shallows and aquatic life, have greatly increased the variety of bird species found in the area.

Other developments, such as new road systems for example, which take place in the vicinity of established shingle structures, frequently result in vehicle movement outside the normal working area. This often results in the wholesale damage of the shingle flora. Regular traffic will soon destroy whole ridge systems; parts of Dungeness and Orfordness have suffered in this way, mainly through the constructional activities of the Ministry of Defence. It could be said, however, that some balance is restored, since such areas are not open to recreational use by the public.

Vegetated shingle structures and beaches are certainly not common in the British Isles. They therefore provide a habitat for particular species of plants which occur nowhere else. This fact, together with the danger to ground-nesting birds from visitors, may require more areas to be protected during

Shingle flora at 'The Crumbles', east of Eastbourne, Sussex.

in their quest for food. As the tide recedes, the sand-inhabitants burrow deeply to avoid not only predators but also the drying effects of both wind and sun.

Indeed, the amount of wind exposure a sandy beach receives is a vital factor in determining both the amount and variety of fauna prepared to inhabit the shoreline. Should the wind be blowing out to sea across an expansive open shore, the waves tend to break early, usually with such violence that the resulting turbulence shifts the coarser grains up the shore. As a result, a gradual gradation of the sand particles occurs, with the finer material being left near the low-water mark. This shifting sand is not conducive to a rich fauna and frequently necessitates the construction of breakwaters to stem the movement.

Where waves continually pound exposed stretches

Breakwaters, in the form of low wooden fences, are erected in several areas, especially coastal resorts, to reduce erosion and keep a beach 'where it is wanted'.

the nesting season. Several sites have been designated National Nature Reserves or Local Nature Reserves by the various conservation bodies, and are strictly wardened during the summer months. This is indeed essential if inadvertent trampling is to be avoided, with the subsequent destruction of nests and eggs. Control of access to shingle sites and limited use by visitors must be regarded as fundamental necessities in the future, if our major shingle formations are to survive. Where wardening is unavailable, the adequate erection of warning notices to encourage a more caring attitude from the public should help to reduce the most damaging effects of trampling.

Much of the coast of Britain is, however, sandy, and to the casual observer a sandy beach seems to be no more than a mass of lifeless sand, with few, if any, signs of plant or animal life. To some extent, this is substantially true, but the situation changes considerably at high water, when all sorts of animals venture out of their hideouts beneath the sand

Sea rocket (*Cakile maritima*) growing on sand at high-tide level on Brancaster Beach, North Norfolk. In the background there is evidence of dune stabilisation using wire mesh.

of beach, the sand is lifted into suspension and individual particles interact, creating a hostile environment which few creatures are able to tolerate. In more sheltered locations, however, the destructive effects of the wave-action are not so pronounced, and numerous microscopic animals are able to survive between the sand particles in association with diatoms and other small algae. Here, too, other edible debris held in suspension by the sea can settle into the sand structure, where it will be broken down by bacteria and fungi to provide food for minute organisms invisible to the naked eye, as well as the larger sediment feeders.

There exists an upper limit to where sand-dwelling animals prefer to live. The coarser grains deposited higher up the shore dry out quickly as the tide goes out. During neap tides, characteristic of spring, the surface sand particles are also liable to wind erosion. Together, these factors make the upper shore inhospitable to sand-dwelling creatures, which thus tend to colonise areas lower down the shore.

To the naked eye, the first noticeable change in structure of a sandy beach occurs at the strandline.

At first glance, it may appear to be no more than a continuous accumulation of wave-driven rubbish and seaweed running parallel to the water line. There is no denying that much of the stranded material is indeed rubbish, with an unfortunate amount of non-biodegradable material in the form of plastic waste in line with present-day packaging. What does decay, however, provides the essential nutrients necessary to sustain a noticeable flora, so that the strandline is about the only place flowering plants encroach onto, in an otherwise barren environment.

Most of the creatures found on the driftline are already dead or certainly in the process of dying. Frequently, some stand out amongst the others by virtue of their relative size; dead birds are a regular sight on our shores, with intolerable numbers being

Marram grass (*Ammophila arenaria*).

Restharrow (*Ononis repens*) and lady's bedstraw (*Galium verum*) growing on a sand dune.

found on areas of coastline susceptible to oil-contamination.

On some beaches, empty shells abound; these will eventually erode to become part of the sand structure. Other shells found still have their occupiers intact inside, but will probably be preyed upon by scavenging birds, many of which look upon the strandline as the supplier of a major part of their diet.

Exposure to wind is an inevitable phenomenon at the coast and, although shingle is hardly affected, sandy beaches are very susceptible. A wind speed of just 16 kph (10 mph) is enough to shift sand particles and on warmer days, when the top layer of sand dries quickly, strong winds will move appreciable amounts across the shore. If the wind speed at ground level falls below the 16 kph (10 mph) level, the sand is deposited. Such an effect is accelerated by debris and plants, which the moving particles first encounter at the strandline. So it is here that the first stages of sand-dune formations occur.

Several factors affect the size and shape of the accumulations formed, including the rate of supply of wind-blown sand and the profile of the shore. The plants forming the barrier are also important, like sea rocket, which is a typical strandline species, and the two grasses (sand couch and lyme), which are even more effective as wind-barriers. The grasses also stabilise developing systems because of their extensive root systems. But there comes a stage when the rate of upward growth of these grasses is not enough to cope with the accretion of sand. An accumulation of more than 30 cm (12 in) a year will have this effect, so that there is a limit to the size of these 'embryo' dunes.

The formation of larger dunes is mainly the result of the presence of marram grass, which is able to keep pace with the rate of sand deposition. A growth rate of 1 m (3 ft) per year is typical in this species and, together with its spreading root system, marram grass is an excellent stabiliser, explaining why it is frequently transplanted to areas where erosion has taken place.

As the dune grows, the top layers of sand come under the influence of a faster wind, which comes along from well above shore level and is therefore not slowed down by surface sand on the beach. The resulting effect is that sand eroded off the top is deposited on the sheltered side of the dune, causing part or sometimes all of the dune to move landwards. There exists a balance between the rate at which the sand blows off the top and the rate at which it blows off the shore. The result is that a maximum height is attained which varies from region to region. Some of the famous dunes at Culbin Sands on the east coast of Scotland are 30 m

A dune slack – a damp area in between stable dunes where vegetation is lush and green.

(100 ft) high, whereas across on the Western Isles flat sandy plains little above sea-level are more usual.

When windspeeds fall, but a good supply of sand still comes off the beach, further developments occur on the seaward side as sand builds up again around those plants responsible for the initial accretion. The end result of all this over a matter of time is the build-up of a series of parallel dunes. Also, the dunes furthest inland do not encounter such strong winds as those in the 'front line', and so further stabilisation occurs as plants colonise them.

There are occasions when the prevailing winds blow offshore. Then, the winds rework material deposited onshore during storm conditions, moving the dune system seawards.

Some of our larger sand dune systems have developed over several thousand years. A typical example would be composed of dry, sandy ridges with occasional flat, damp plains known as 'slacks' in between. Good examples occur all round the British coastline, but in particular Norfolk, Wales and north-east Scotland.

Of all the coastal habitats, however, sandy shores are the most susceptible to human disturbance, which at times can be overwhelming. In many cases, especially in England and Wales, dune systems develop behind popular beaches, and their use for recreation poses a significant threat to their future. Erosion, caused by trampling, motorcycling, horse-riding and in some instances camping, serves to break up the vegetation. This exposes the sand to the wind, and without some form of artificial stabilisation leads to so-called 'blow-outs' where substantial areas of sand dune may be lost.

This means that good management is essential,

Boardwalks being used to reduce dune-erosion by visitors.

Dune repair.

and it is heartwarming to note that this is being carried out all over the country by the various conservation bodies, both national and local. To reduce human disturbance, it is necessary to fence off certain areas under threat and to provide boardwalks to the beach. A more direct measure is to erect windbreaks to trap sand, which together with the planting of marram grass will help to reverse the process of dune-loss.

Other threats to the existence of dune systems rich in wildlife tend to be localised. During the last century, afforestation has been initiated in several places to prevent large-scale movement of sand, such as at Newborough in Anglesey and Tentsmuir in Fife. The practice has certainly achieved its aims, but in doing so has changed the character of the areas as far as the diversity of wildlife they contain is concerned.

It would be natural to assume that the use of coastal dunes for golf courses would result in a substantial decrease in wildlife. This is not always the case, however, and although existing vegetation is destroyed by the construction of fairways, bunkers and greens the surrounding area of rough ground provides a refuge for many of the dune plants and animals of the original habitat. In fact, the roughs at Royal St George's golf course at Sandwich in Kent support the biggest colony of the rare lizard orchid in Britain.

Changes in agriculture over the years have meant a decline in animal grazing, allowing coarser scrub to dominate in areas which were once botanically rich. At several major sites, scrub, especially sea buckthorn, has been eradicated from certain areas to attract a wider variety of wildlife.

In the Western Isles of Scotland, government and EEC plans for the possible intensification of farming on the sandy plains known as the machair pose a significant threat to this habitat.

Sand dunes are vital habitat for a rich variety of plants and animals. They are also of infinite importance to our coastal defence structure. It is imperative that they are preserved, not just by the individuals responsible for their management, but by the whole nation.

Flora and Fauna

Although large stretches of shingle are usually too mobile to support plant life, where a build-up of humus occurs along the strandline there is relative stability allowing plants such as sea beet and common orache to germinate and grow. Just above the strandline, where wind-blown sand accumulates, sea sandwort, sea couch and red fescue may also colonise the shingle. On Scottish coasts, the rare oysterplant frequently forms a mat of vegetation on shingle beaches, as does the sea pea, which is mainly confined to eastern and southern coasts. Decaying sea pea plants greatly enrich the nitrogen content of the driftline humus since their roots bear nodules, containing nitrogen-fixing bacteria which are able to absorb atmospheric nitrogen. Where the humus has built up on stable shingle, a more closed vegetation develops and the familiar flowers of sea campion and sea thrift carpet the floor. The low-lying campion foliage also tends to form a mat over the shingle, thus acting as a stabiliser.

In exposed shingle environments, such as at Orfordness in Suffolk, plants occur which are specially adapted to such conditions. These include the yellow horned-poppy and the edible sea kale, which belongs to the cabbage family. Their survival is very much dependent on there being some sand or soil between the pebbles, into which they can extend their long roots. Both species are able to survive the occasional movement of the shingle during storms.

Undisturbed exposed shingle environments soon develop a rich lichen-dominated community and the survival of these fragile slow-growing plants depends a great deal on the area being totally protected from human agitation, as is Orfordness – a designated National Nature Reserve.

In sheltered regions, such as along the lee side of shingle spits, the driftline may become colonised locally by shrubby seablite, although the plant is confined to southern Britain, in particular the shingle shores of north Norfolk. Scrub only develops on the least disturbed stable shingle and is usually limited to gorse and bramble. Nevertheless, individual stunted trees frequently survive as at Dungeness in Kent, where single wind-lashed bushes occur on otherwise bare shingle.

Several of the plants common on sandy shores may be found on shingle beaches which have a sand content, particularly the spiny perennial sea holly, sea bindweed and sea rocket.

Stable shingle is of considerable importance to nesting birds such as ringed plover and oystercatcher. On undisturbed shingle sites, common terns, arctic terns and sandwich terns frequently nest in large colonies. Their nest is no more than a depression in the ground, as is that of the little tern, which is the rarest of our breeding terns. Occasionally, the common gull, herring gull and lesser black-backed gull will choose shingle as a nest site.

Small migrant birds use the scrub on some shingle sites as important nesting sites or feeding areas during passage. In winter, turnstones feed amongst the pebbles of the shingle shore. The commonest mammals of shingle shores are rabbits and hares. On some undisturbed sites, grey seals come ashore to rest or make use of the tranquillity to give birth to their young.

On a predominantly sandy shore, the only plants able to survive the pounding waves are small algae which attach themselves to the sand particles. Since they need light for photosynthesis, they are found in the upper layers of sand, no more than a few millimetres below the surface.

Animal life is much more abundant, although this may not be altogether apparent to the casual ob-

Sea buckthorn (*Hippophaë rhamnoides*) in berry on fixed sand dunes.

server. The larger animals of sheltered beaches can be arranged into three groups – deposit-feeders, suspension-feeders and carnivores, including omnivorous scavengers.

Deposit-feeders include those animals which feed on the material that settles in the sand as the water passes over it. They form a large group and, like most of the resident animals of the sandy shore, live in the proximity of the low-water mark. One exception is the lugworm, which is found under most of the exposed sand. Other species in the group include the sea potato whose empty skeletons can be found on the beach after rough seas, and also the sand-mason and some cat-worms.

Suspension-feeders filter suspended material in the water when covered by the tide. The majority are known as bivalve molluscs and include cockles and razorshells.

A habitat is not complete without its resident scavengers, and one animal which does its job well is the sand-burrowing starfish, which tends to eat most things it comes across. Other omnivorous scavengers include shrimps and the burrowing masked crab.

Other non-resident, but visible, scavengers also visit the beach at different times of the day depending on the state of the tide. Although not entirely restricted to sandy shores, wading birds probe the sand surface with their long bills in search of food. By far the most commonly seen is the oystercatcher, very partial to cockles in particular, although it will also eat catworms and lugworms in addition to other bivalves.

Near the strandline, the tiny holes in the sand are the burrows of sand-hoppers, which surface at night to feed on any available plant or animal material amongst the debris. In the strandline itself, most of the animals found are already dead or in the process of dying. Typical victims of seasonal storms include masked and common swimming crabs, razorshells and starfish. Other less obvious items commonly found include the egg-cases of whelks, skate and dogfish.

At the highest strandline, the first noticeable plants occur, with such species as sea rocket, common orache and prickly saltwort forming the first real barrier to wind-blown sand. Beyond this level, small accretions of sand form what are called fore-dunes, on which sand couch and lyme grass grow. Both are able to tolerate the occasional immersion in seawater during the highest spring tides.

The accumulations of sand become much more extensive further from the shore. Here marram grass is the dominant stabiliser and dunes are often referred to as 'yellow dunes' or even 'white dunes', due to the colour of the clearly visible, uncolonised, bare sand. Most plants growing in these 'front-line' dunes are able to put up with the deficiency of water, plant nutrients, exposure to strong winds and the high temperature of the upper sand layers in strong

sunshine. Apart from the dune-builders, few plants are able to tolerate this hostile environment except for such species as dune fescue, sand sedge, sea bindweed, sea spurge and sea holly, which flourish in many localities.

Behind the first layers or dunes, sand stability increases, and, although marram grass is still present, a more diverse flora survives. Lichens carpet areas of bare ground, producing a greyish colour. At this stage, the dunes are sometimes referred to as 'grey dunes'. Other plant species colonising the region vary depending on the type of sand present. Where the sand grains are mainly made up of calcareous material, including seashell fragments, dune helleborine, ragwort and viper's bugloss may flourish, with marsh helleborine and creeping willow confined to the damper dune-slacks.

In some areas, where the dunes are made up of silica sand, the soil is more acidic, allowing such acid-loving plants as heather to grow.

Many dune systems are heavily populated with rabbits, whose constant grazing may have some effect on the plant species present. Systematic grazing results in a lichen-rich flora, but where rabbits are few in number, or indeed absent, a dense scrub of sea buckthorn or hawthorn soon takes over. Few of the smaller plant species can survive in this situation, but at least the scrub does provide good breeding and feeding areas for songbirds.

Of the seabirds nesting on the dunes themselves, arctic, common and sandwich terns are amongst the 'regulars'. Other birds nesting on the ground include meadow pipits and skylarks, but in well-concealed locations. Adders are common in some areas and a constant worry to the public, especially parents with young children playing away from the shore. A number of dune systems throughout England and Wales support important populations of the rare natterjack toad, which requires a fine balance of wet slacks and sandy area to survive.

Insect life can be very rich at times, with cinnabar moths present in numbers where ragwort flourishes. It is the black and yellow larvae that are most often seen feeding on the plant. Of the butterflies, it is possible to see a variety of inland species at the coast, especially the common blue. The grayling and dark green fritillary are also particularly fond of the dune habitat and are often seen in the same locations during August, chiefly on western coasts.

Common Gull (*Larus canus*)

Length 41 cm (16 in)

A small gull, similar in length to the kittiwake (*Rissa tridactyla*) but slightly stouter; as they are unlikely to be in company with each other the chance of confusion is remote. Unfortunately the name leads one to suspect it may occur as a common species throughout Britain, but this is not the case. In fact it is the least numerous of all our coastal-breeding gulls. The American name 'mew gull' would certainly be more appropriate. At a distance, and with no means of size comparison, differentiation between the herring gull (*Larus argentatus*) (which at 56 cm, 22 in, is much larger) and the common gull may cause minor problems. In close-up, the darker eyes, the greenish-yellow bill, and legs of the common are diagnostic.

During the winter months it is a rather silent bird, feeding on pasture and arable land by day, and roosting on estuaries or large inland waters at night. At this time of year their numbers are swelled by visitors from Scandinavia and the Baltic. Surprisingly few common gulls ringed as nestlings in this country have been recovered abroad.

The gulls begin returning to their breeding grounds (moors, hillsides and around lochs etc) as early as February and March. The males are intent on establishing territories, which they guard against rival males by assuming an upright threat posture with wings slightly raised. Females are permitted entry to their territory until one has been chosen as a mate. Then they become very vocal, uttering shrill cries of 'kyow' or 'kik-kik-kik', as they angrily intercept any intruders. 'Head-flagging' is another common feature at this time.

Nest-building soon follows when a shallow depression in the ground is sparsely lined with vegetation gathered close at hand and small nesting colonies begin to form. (Occasionally nests are constructed on the tops of bushes and in trees.) At the end of May or early in June egg-laying begins; the normal clutch is three, the background colour varying from pale blue to some shade of green, blotched and spotted with dark browns and greys. Eggs are laid at intervals of two days; incubation commences with the last egg and is undertaken by both birds. The downy grey-brown chicks hatch after a period of 24 to 28 days and leave the nest within three or four days. They remain in the vicinity until fledged; this takes a further five or six weeks.

Common gull (*Larus canus*) and chicks.

Common gulls feed by foraging in fields and wet pastures, and on moorland. They search for earthworms and a variety of other invertebrates; grain is also taken. The shoreline is often visited when small items of marine life are taken.

Breeding distribution: Iceland, Faeroe Islands, Ireland, Scotland, and a few localities in England and Wales. Westwards between 50° and 70° and into Asia, also Alaska and north-east Canada.

Ringed Plover (*Charadrius hiaticula*)
Length 19 cm (7½ in)

A very widespread species typical of our low-lying shores. Upperparts mainly sandy brown, underparts white. It has a black band on front of crown; the white band on forehead extends behind the eye. A dark band continues from base of bill backwards through and below the eye. There is a white upper ring and a broader black lower ring around the neck. Bill orange with black tip; legs and feet orange. In flight prominent white wing bar distinguishes it from the little ringed plover (*C. dubius*).

The call is a musical piping 'too-li', the song an oft-repeated 'quito-weeoo' which gradually increases in tempo until it becomes a trill.

Ringed plover (*Charadrius hiaticula*).

It breeds on shingle or sandy shores, also on closely-cropped maritime turf. The ringed plover can be very vulnerable, breeding or attempting to breed as it still does in traditional sites which are now the playgrounds of holidaymakers.

Egg-laying commences in April, continuing through to June or even later. A shallow scrape in the sand or shingle serves as a nest; this is lined with a few tiny pebbles or shell fragments into which four eggs are laid. These are buff-coloured, spotted with browns and black. They are often very difficult to detect on a shingle beach. Incubation lasts for 22 to 24 days and is shared by both sexes. Should a human approach the nesting area the sitting bird quickly leaves the nest and very often attempts to lure the intruder away by running around with widespread tail, and a trailing wing as if it were broken.

When feeding, the ringed plover runs along, stopping abruptly and tilting the entire body forward as it picks up some choice item. Often seen at the water's edge in company with other small waders, especially dunlin (*Calidris alpina*).

It is of course not confined as a breeding species to coastal areas but can be found inland on the shingle banks of rivers and lakes, as well as gravel pits. In East Anglia many nests have been found in fields of sugar beet.

Its breeding distribution includes Britain, parts of western Europe and Scandinavia; with southern Europe being the principal wintering area, many birds migrating to southern France, Spain and Portugal.

In the Arctic tundra regions of Europe and Asia it is replaced by the somewhat darker sub-species *C.h. tundrae*.

Lesser Black-backed Gull (*Larus fuscus*)
Length 53–56 cm (21–22 in)

A medium-sized gull, the adult birds of which have back and wings slate-grey; the black tips of primaries are marked with white. Rest of plumage white. Bill yellow with a red spot towards tip of lower mandible. The yellow legs help distinguish it from the slightly larger herring gull (*L. argentatus*). Less numerous as a breeding species in Britain than the herring gull, with 50,000 pairs as against 300,000. The densest populations occur in Scotland (the Clyde and Forth), south-west England and

Lesser black-backed gull (*Larus fuscus*) with chicks, on Walney Island.

South Wales; whilst the largest colony (which accounts for almost one-third the total in the British Isles) is interspersed with breeding herring gulls on the sand dunes of Walney Island (near Barrow-in-Furness). Although a small number of lesser black-backs do remain on our shores throughout the entire year, it is essentially a migratory species, travelling to the warmer climes of Africa and the Mediterranean in autumn.

March is the month when peak numbers return to their breeding grounds, and nest-building usually commences towards the end of April, with eggs being laid May/June. A normal clutch would be three eggs; these vary widely in ground colour, ranging from pale greenish-blue to dark olive-green, spotted and streaked with blackish-brown. Incubation is undertaken by both sexes for a period of 28 days, after which the chicks stay close to their nest. Those which inadvertently wander away are likely to be killed by other adults. Interbreeding between the lesser black-backed and the herring gull has been recorded on but a few occasions.

During the breeding season lesser black-backs do not scavenge for food as do the herring gulls and great black-backs (*L. marinus*). They do, however, prey on other birds and will take their eggs; but fish, marine worms, and crustaceans make up their diet, supplemented by carrion and occasionally grain. However, in winter months some now join forces with other gulls and have become dependent on the scavenging method, visiting refuse tips close to large towns.

In Northern Europe there are two distinct races; the darker-backed nominate *fuscus* (the Scandinavian lesser black-backed), and the British race *graellsii*. Their breeding range extends south from Iceland, the Faeroe Islands, Britain, Ireland, and southwards (though intermittently) from Scandinavia and the Baltic to Brittany. The pink-legged form occurs in the Kola Peninsula of Russia.

Arctic Tern (*Sterna paradisaea*)
Length 36–38 cm (14–15 in) **Fig. 6(1)**

The differences between the arctic and the common tern (*S. hirunda*) are indeed slight. Both are pale grey above and white below. During summer the cap and nape are black, the forehead and crown becoming mottled with white in winter. The arctic tern has a completely red bill in summer (with few exceptions; however, some do retain a blackish tip), whereas the common tern has a distinct black tip to its bill. Both have red legs. During winter the bills of each are black with a reddish base. There are a few distinguishing features. In silhouette the extention of the arctic's head is not far beyond its wings, the red bill and red legs are just a little shorter. When the wing is folded the tail streamers project slightly further than the wing tips, and when in flight the primaries appear almost translucent.

The common may be the most widespread of the terns breeding in Britain, but the arctic is certainly the most numerous with a possible population in the region of 77,000 pairs; and approximately 90 per cent of these being just about equally divided between Orkney and Shetland.

British breeders seem to have a diet consisting mainly of sand eels, sprats, and small herrings, supplemented by small crustaceans. Arctic terns are colonial nesters, choosing sandbanks, shingle, and open ground close to the shore. They return to their breeding grounds towards the end of April or early May; at this time fish-carrying flights play an important role, followed by the presentation of fish to a would-be mate. Other displays include greeting ceremonies between the sexes when head and bill are held vertically erect, or pointing acutely downwards with head hunched in.

The nest is a mere scrape, sparsely lined with dry grasses or a few tiny pebbles and fragmented sea shells. It is usually quite open, although on occasions may be well concealed under vegetation. During late May or early June one to three eggs are laid, and the colour variation that exists between clutches, or even between eggs within a clutch, defies description. They are quite indistinguishable from those of the common tern, and receive equal incubation from both birds for a period of 22 days. The chicks are fed on average about 25 times per day, and fledge within three or four weeks.

Terns are notorious for their swooping attacks on all intruders and the arctic is regarded as the most aggressive. It is also one of the world's great bird-migrants, some travelling a round trip of over 32,000 km (20,000 miles) or more as they journey to their wintering grounds in the Antarctic before returning to their Arctic breeding grounds. Distribution as a breeding species in Europe and Asia: Iceland, Faeroe Islands, N.W. Russia, Britain, Ireland, the coasts of the North Sea and the Baltic, northern Scandinavia and Brittany.

Little Tern (*Sterna albifrons*)
Length 23–25 cm (9–10 in) **Fig. 6(4)**

As its name may suggest, this is the smallest of the terns to breed in Britain and Ireland. Its mantle, rump and dark-tipped wings are pale grey; the crown and nape are black, as is a line which extends through the dark eye to base of bill. Rest of plumage white; retains its white forehead at all times, the area of white becoming broader during winter months. Bill slender and yellow with pointed black tip. Legs and feet are a duller shade of yellow than the bill. Juveniles are similar with that typical forked tail, but head and back are speckled darkish brown, forehead ashy-grey, legs and bill are brownish yellow. They attain adult plumage by the first summer.

The little tern nests in small scattered colonies, and of the 2,000 or so breeding pairs in the British Isles the coastal areas of East Anglia and Lincolnshire account for about 65 per cent. In recent years their numbers have declined alarmingly, partly due to the encroachment of holiday-makers onto sandy/shingle beaches which for centuries had provided undisturbed conditions so essential for successful and continued breeding. Fortunately many of the small colonies have now been fenced off. Predation by foxes and the effects of high spring tides have also taken their toll.

It is during April and early May that the little terns return to the vicinity of their breeding sites. Both birds help towards fashioning a scrape in the sand or dry mud; these are usually left unlined, but nests constructed on shingle beaches are defined by the addition of tiny pebbles. Egg-laying commences during May when two or three stone-coloured or pale blue eggs are laid; these are finely spotted with varying shades of brown and exhibit all manner of ashy-grey markings. Incubation commences with the second egg, a duty shared by both birds and continued for about 21 days. An excited cry of

Fig. 6   1. Arctic tern (*Sterna paradisaea*). 2. Common tern (*Sterna hirundo*). 3. Sandwich tern (*Thalasseus sandvicensis*). 4. Little tern (*Sterna albifrons*). 5. Roseate tern (*Sterna dougallii*).

'kirri-kirri' can be heard as the birds display over their breeding territory.

The diet varies distinctly, from one consisting almost entirely of sand eels on the one hand to one comprised mainly of crustacea and marine worms on the other. Foraging birds utter a high pitched 'kik-kik-kik'.

On hatching, the chicks are cared for by both parents and make their first attempts to fly within 28 days. Vacation of the colonies commences towards the end of July.

In Northern Europe the little tern does not occur in Shetland or Orkney, nor the coast of Belgium; but otherwise can be found around practically the entire coastline of Europe; northern Scandinavia being the only other exception. It extends eastwards through central Asia to Afghanistan and western Altai, then south to the Persian Gulf and along the west coast of India. Breeding also takes place in the Far East, Australia, Africa, both coasts of the United States, West Indies, and on islands off Venezuela. Thus it has the widest breeding range of all the British terns.

Common Tern (*Sterna hirundo*)
Length 35 cm (14 in) **Fig. 6(2)**

In the British Isles, the population of the common tern is put at 15,000 pairs; it closely resembles the arctic tern (*Sterna paradisae*) (whose estimated population is thought to be in excess of 75,000 pairs). Both are pale grey above, and white below; in summer, cap and nape are black; the bill orange-red with a distinct black tip (that of the Arctic tern is entirely red, although a few individuals do have a slightly dusky tip). They both have red legs, and if standing side by side it would be seen that the common tern's legs are slightly longer. Look also for tail streamers which project a little way beyond the tips of closed wing in the Arctic tern.

Almost half the entire British population can be accounted for in just four coastal sites; one at Strangford Lough in Northern Ireland; one in Norfolk; another in Northumbria; and the fourth site in Shetland. There are scattered colonies widely distributed around the coast except in South Wales and south-west England (but including the Isles of Scilly). Inland sites occur principally in eastern England and north-east Scotland. Listen for the call, which is a sharp 'kik-kik-kik-keerr-keerr'.

Nesting-sites vary a great deal and include both sandy and shingle beaches, small islands off-shore, saltmarshes, and sand dunes. Common terns nest inland more frequently than other terns, often choosing gravel spits in rivers, freshwater marshes, also sand and gravel pits. The nest may be a simple ground scrape sparsely lined with a little dry vegetation, or the eggs may be laid on bare rock. An average-sized clutch would be two or three eggs, the ground colour in some pale shade of blue, green, brown, or cream; they are usually boldly marked with dark browns and greys. Laying is from mid-May onwards; the eggs are incubated alternately by both sexes for 22 to 26 days. Both parents help to feed the chicks which fledge after about four weeks. Sand eels and other small fish form the bulk of the diet, and are caught by plunge-diving; crustaceans, molluscs and marine worms supplement this diet.

The main wintering quarters for European breeders are in West Africa. North American birds journey to South America.

S.h. hirundo (the nominate) breeds in Europe, W. Asia, and N. America. There are three sub-species: *S.h. betana* of Turkestan and Tibet; *S.h. minussensis*, C. Asia, N. Mongolia; *S.h. longipennis*, N.E. Asia.

Roseate Tern (*Sterna dougallii*)
Length 38 cm (15 in) **Fig. 6(5)**

Nowadays, with possibly less than 1,000 breeding pairs, this is certainly the rarest British tern. Since this small British population represents about 80 per cent of the total for the whole of north-western Europe, the roseate tern's status there can only be described as precarious.

At long range it is easily confused with both the arctic (*S. paradisaea*) and the common (*S. hirundo*). There are, however, some fundamental differences in adult birds; for instance the red legs are noticeably longer; the predominantly black bill (slightly red at base when breeding) is finer and somewhat longer. Upperparts are of a much paler grey, and the white breast is suffused with pink when breeding (but this can only be seen at close quarters). When in flight amongst Arctic and common terns, look for the longer tail streamers and much whiter body.

Roseates usually nest in small groups amongst other terns or at the edge of a colony. The nest can be well hidden in thick vegetation, or it may be just a simple scrape without lining. A clutch of two eggs is normal, and laying begins towards the end of May. The cream or buff eggs are slightly elongated, streaked and spotted with brown, and have faint underlying markings of ashy-grey. Both birds share the 21 to 26 days' incubation, and help feed the young, which leave the nest after a few days but do not fledge for a further 21 to 28 days. Sprats and sand eels provide the bulk of the roseate tern's diet.

Listen for the guttural alarm note 'aak, aak'; and when carrying fish a soft 'chevick'.

It nests mainly on small rocky or sandy islands. Northumberland is the only eastern county where it nests regularly, for example the Farne Islands. There is a colony of note on Anglesey, Wales; but the largest concentrations occur on three sites in Ireland, namely Green Island, Strangford Lough, and Lady Island Lake.

Vacation of the colonies begins at the end of July; roseates begin to return in the following April.

S.d. dougallii (the nominate) breeds in Britain, Ireland and Brittany (occasionally W. Germany and S. France), the Atlantic coast of N. America from Nova Scotia down to C. America, the West Indies, and islands off the coast of Venezuela, also the Azores and islands off West Africa. There are four sub-species: *S.d. korustes*, Sri Lanka, Andaman Is; *S.d. arideensis*, Seychelles, Mascarene Is; *S.d. bangsi*, Riuku Is, Philippine Is, Kei Is, and Solomon Is; *S.d. gracilis*, Moluccas, N. and W. coasts of Australia.

Turnstone (*Arenaria interpres*)
Length 23 cm (9 in) **Fig. 7(1)**

The turnstone is primarily an arctic breeding wader whose range extends southwards around the coastal areas of Scandinavia. Although we have on occasions encountered these birds (apparently paired) in the Outer Hebrides during the month of May, there was no positive evidence of breeding. We suspect others will have had similar sightings. So for Britain it must be regarded as just a winter visitor.

Its winter habitat is principally rocky coasts and along sandy or muddy shorelines where it can be seen in small parties searching amongst pebbles or upturning shells and small stones for tiny crustaceans and small molluscs. On arrival to Britain some will still retain their distinctive summer or transitional plumage. Head, neck and tail being

patterned in black and white; underparts white; with remainder of upperparts patterned in reddish-brown and black; producing a 'tortoiseshell' effect. Those already in winter dress are much more sombre; gone is that familiar pattern. Now the turnstone has upperparts of a more uniform dark brown, underparts white, but legs still orangey-red.

In flight the white rump exhibits a black crescent-shaped patch; the white-tipped tail has a distinctive broad black terminal band, also white shoulder patches become apparent.

The nominate *A.i. interpres* has a breeding range which extends from Ellesmere Island (Arctic Canada) and south-west Greenland eastwards to Iceland (where it breeds occasionally), Denmark, Norway, Sweden, Finland and throughout the Arctic from Novaya Zemlya to the Chuckchi Peninsula. A sub-species *A.i. morinella*, the ruddy turnstone, breeds in Alaska and Arctic Canada east to Baffin Island.

Good areas for winter watching include Morecambe Bay, Lindisfarne (Holy Island), the Wash, Dengie in Essex, Burrey Inlet in Glamorgan, and the Firth of Forth.

Black-headed Gull (*Larus ridibundus*)
Length 36–38 cm (14–15 in)

The smallest of Britain's resident gulls, and the most widely distributed inland breeder; with an estimated population in excess of 250,000 birds. In summer easily distinguished by its chocolate-brown head, and the whitish ring around the eye. Bill and legs are bright red. In winter it loses the dark cap, and there is then a blackish smudge behind the eye. The grey wings retain their black tip in all seasons. Look for the characteristic long wedge of white on primaries when in flight. Rest of body plumage remains white. The only possible confusion in spring and summer could be with the much smaller little gull (*Larus minutus*) (length 28 cm, 11 in); look for jet black cap which extends down the nape, also rather rounded wing tips. Also with the larger mediterranean gull *Larus melanocephalus* (length 40 cm, 16 in). Look for lack of black tips to white primaries; black cap extending well down the nape; dark band crossing the red bill; the white ring around the eye is not entire.

The black-headed gull is very common on the East Anglian coast. Absent as a coastal breeder in

Fig. 7 1. Turnstones (*Arenaria interpres*). 2. Purple sandpiper (*Calidris maritima*). Birds in winter plumage.

south-west England and most of Wales. A survey of coastal areas during 1969–70 put the population at about 74,500 pairs; the remainder can be regarded as inland breeders.

A colonial nester, the range of sites includes both freshwater and saltmarshes, shingle banks, coastal lagoons, sand dunes, moorland, and even sewage farms. They re-form their colonies during March or early April. 'Neck-stretching' and 'head-flagging' form part of their display. Nests may be a shallow scrape on dry land and only sparsely lined; but those constructed in wet areas are much more substantial and often a bulky collection of dead vegetation; when in marshes, grassy hummocks and dense tufts of vegetation are utilised to keep the nest above water level. A normal clutch would be three eggs; the ground colour may vary between shades of brown, green or blue, but always covered with dark brown and grey markings. Both birds share the 22 to 24 days period of incubation. The young are fed by both parents, and fledge after five or six weeks.

Two coastal colonies of note are at Needs Oar Point, Hampshire; and Ravenglass, Cumbria.

World distribution includes Europe, Asia to N. Africa, India, and the Phillippine Is. There are no recognised sub-species throughout this range.

Sandwich Tern (*Thalasseus sandvicensis*)
Length 38–43 cm (15–17 in) **Fig. 6(3)**
Those slender bodies and narrow pointed wings make terns easy to distinguish from gulls. The sandwich tern is the largest and heaviest tern to breed in Britain and Ireland; it has a relatively short tail, so consequently is the most gull-like in appearance.

During summer the elongated feathers of the black crown are raised to form a crest when the bird becomes excited. The back and wings are pale silvery-grey, remaining body plumage is white. It has a yellow tip to the black bill. Legs and feet are black. Its loud call of 'kirrick' is far-carrying and often the first indication of its presence.

The total British population is put at a little over 15,000 pairs. The sandwich tern breeds almost exclusively in coastal areas, the largest numbers occur in Norfolk and Northumberland. Apart from Anglesey it is absent from Wales; nor does it breed in Shetland; and in south-west England can only be found in Dorset.

Bill-raising with neck outstretched and wings half open is included in the display; there is also much bowing, and bill-to-bill contact. Colonies are formed during April and nest-sites are occupied later that month or early in May. Sand dunes, shingle banks, and sometimes low grassy islands, are typical sites. A simple scrape or depression serves as the nest, into which two eggs are laid; they are usually some shade of buff and exhibit a variety of markings, reddish-brown smears and spots, and a few grey scrawls. Both birds share the incubation for a period of three to four weeks, and the young fledge when about four weeks old. The food in summer consists mainly of sand eels and sprats which the terns acquire by plunge-diving. The young are dependent on their parents to some degree for the first three or four months.

T.s. sandvicensis (the nominate) breeds in W. and S. Europe, and winters in Africa and N.W. India; the only sub-species *T.s. acuflavidus* occurs in Florida and the Gulf coast.

Sandwich tern colonies are often not very static from one year to the next, and their numbers may fluctuate alarmingly.

Shore Lark (*Eremophila alpestris*)
Length 16.5 cm (6½ in) **Fig. 1(2)**
The northern European race *E.a. flava* breeds in the Arctic tundra. It winters, as its name implies, along the shoreline. A gregarious species usually seen in small flocks and often in company with groups of snow bunting (*Plectrophenax nivalis*).

The male in summer has sides of crown, cheeks, and upper breast black; rest of face and throat are yellow; this striking facial pattern is topped with a small black erectile horn on either side of the crown, hence the American name of horned lark. Wings and upperparts are pinkish-brown graduating to white on the underparts. The female has no horns, and her head pattern is far less distinct; rest of plumage similar to the male's.

Most of the northern European birds winter along the sandy shores and beaches of the Baltic and North Sea. On passage the shore lark visits the east coast of England regularly, albeit in small numbers. Then the male does not have the two black horns.

The world distribution includes 41 races which cover most of North America; Mexico, and into Colombia; N. Europe; Asia to China (including the Himalayas); Asia Minor; Iran; Lebanon; and Morocco.

List of Plant Species

Sand Sedge (*Carex arenaria*)
Family Cyperaceae (145) **Fig. 8(3)**

Likes a sandy saline soil in which its extensively creeping roots form a matted kind of growth over wide areas. Because of this habit it has been used to arrest the sea's incursion in many coastal areas of Britain, and also planted on dykes in the Netherlands. Sand sedge is a very familiar species and widely distributed on sandy coasts; is only rarely found inland. Its short rough stems are solid, acutely triangular in section, and 10–40 cm (4–16 in) long. The leaves are either equal in length to, or a little shorter than the stems; 1.5–3.5 mm wide and with rolled-back margins. The flowers are borne on spikelets, they are sessile, touching, pale brown, and flattened at margin. Each spike is comprised of between 5 and 12 spikelets, and measures up to 4 cm (1½ in). Terminal spikelets are comprised of barren male flowers, the middle spikelets may have male flowers above and female ones below, whilst the lower ones have only female flowers. The fruit is in the form of a single-seeded nut, and these fall in the immediate vicinity when ripe; because they do not split open to release the seed, so clumps of sand sedge eventually form. However, vegetative reproduction occurs to a great extent. It flowers from June to July and can be found in all our coastal counties, especially on fixed dunes.

Lop-grass (*Bromus mollis* agg.)
Family Gramineae (146) **Fig. 8(6)**

Common throughout the British Isles but more so in the south; in no way confined to the coast but is of widespread occurrence in meadows and on waste land. An annual or biennial of 5 to 80 cm (2–31 in). The culms (grass stalks) usually erect, but diameter varies with the individual; may or may not be covered with short hairs. The panicles are 5–10 cm (2–4 in) long, more or less erect and usually branched; each spikelet-stalk is usually no longer than the spikelet itself and sometimes much shorter. Spikelets are 10–20 mm long, lanceolate, may or may not be hairy, and compressed; never in groups of three at ends of branches. The lower bract (lemma) of each flower is between 6.5 and 9 mm long with prominent nerves, ovate and weakly angled. The upper bract (palea) is more or less the same in length as lower bract. Flowers from May to July on dunes, shingle-banks and cliffs. Occurs commonly throughout the British Isles, less so in north.

Creeping Fescue (*Festuca rubra* var. *arenaria*)
Family Gramineae (146) **Fig. 8(1)**

A very variable but more or less erect perennial of 10–70 cm (4–28 in). Usually stoloniferous (having creeping stems). Panicles erect and somewhat spreading. Spikelets are large and pubescent, each has from four to eight reddish or purple florets. Culm(stalk)-leaves are stiff, flattish, 0.5–3 mm wide, with or without hairs, and blunt or slightly pointed. Leaves of sterile shoots are bristle-like and rolled. In flower from May to July. Grows on sand dunes and is generally distributed in coastal areas throughout the British Isles.

Curled Dock (*Rumex crispus*)
Family Polygonaceae (79) **Fig. 8(2)**

By far the most common species of British dock. It grows from 50 to 100 cm (20–40 in) tall; the habit is erect and somewhat pyramidal. The stem is branched, spreading and often reddish. Its large leaves are up to 30 cm (12 in), usually narrow and

Fig. 8 1. Creeping fescue (*Festuca rubra* var. *arenaria*). 2. Curled dock (*Rumex crispus*). 3. Sand sedge (*Carex arenaria*). 4. Seaside or slender thistle (*Carduus tenuifloris*). 5. Northern shorewort (*Mertensia maritima*). 6. Lop-grass (*Bromus mollis* agg.).

lanceolate, rounded at base, and tapering to an obtuse point from about midway, margins very wavy. The leaf characteristics alone should be sufficient to differentiate the curled dock from any other species of *Rumex*. Flowers greenish, in a panicle; and arranged in whorls which may be close, separate, or united, and towards the end of the branches. They are wind-pollinated. The fruit perianth-segments (separate leaves which comprise the flowers, especially when sepals and petals are indistinguishable) are three in number, they are more or less broadly heart-shaped, and average 4.5 mm. In flower from June to October. Grows on dunes and shingle beaches in coastal areas. However, this very common perennial is also generally distributed in grassy places, cultivated land, and waste places.

Northern Shorewort (*Mertensia maritima*)
Family Boraginaceae (103) **Fig. 8(5)**

As its name suggests, the habitat of this rather fleshy herbaceous perennial is along our northern seashores. The habit is prostrate, then ascending towards end of its leafy and much-branched, purplish stems (which appear to be coated with hoar frost). The plant is hairless, bluish-green, up to 60 cm (24 in), and stoloniferous. Its fleshy, more or less egg-shaped leaves are rough, the upper surface covered with hard dots. They are 2–6 cm ($\frac{3}{4}$–2$\frac{1}{4}$ in) long and arranged in two rows; the lower ones taper to a stalk, the upper ones are stalkless. The flowers are at first pink, later blue and pink; they are in forked cymes and have two leaf-bracts below, set opposite. Corolla is about 6 mm in diameter, five-lobed to the middle, and has yellow folds or protuberances in the throat. Flower-stems 5–10 mm, becoming curved-back in fruit. Its leaves are said to taste like oysters, hence the alternative name of 'oyster plant'. In flower from May to August. Grows locally in Caernarvon, Anglesey, Norfolk, also north Lancashire, and Northumberland to Shetland. Very rare in south. Rare and local in Ireland, where it occurs along east and north coasts, from Wicklow to Donegal. Appears to be on the decrease almost everywhere.

Seaside or Slender Thistle (*Carduus tenuifloris*)
Family Compositae (120) **Fig. 8(4)**

The stem of this annual or biennial thistle is erect and 15–120 cm (6–47 in) tall. A distinguishing feature is provided by the leaves whose bases form a continuous spinous wing along almost the full length of the stem, stopping just short of the flower-heads; basal leaves are more or less lanceolate. All the leaves are somewhat cottony beneath, and pinnatifid with wavy spinous margins. The small cylindrical flowerheads are about 15 × 8 mm and arranged in dense terminal clusters of between three and ten sessile heads (sometimes more). They are comprised of many pale purplish-red florets; the thin dry inner bracts are of equal length or a little longer than florets. The outer-bracts of flowerheads are ovately lance-shaped terminating in an outwardly curved spine, and void of hairs. Each single seed has a long tuft of white down. Flowering period is from June to August. Grows on waste places and waysides especially near the sea. It is locally common in coastal Britain northwards to the Moray Firth and the Clyde; occurs throughout Ireland.

Sea Lyme Grass (*Elymus arenarius*)
Family Gramineae (146) **Fig. 9(4)**

A familiar seashore grass where it grows on sand dunes often in association with marram grass (*Ammophila arenaria*); squirrel-tail grass (*Hordeum marinum*); and in the south of England hedgehog grass (*Cynosurus echinatus*). It is an erect and robust perennial 1 to 2 m (3–6 ft) tall, and propagated by soboles (underground trailing shoots). Sea lyme grass is wind-pollinated; the flowers are in a dense upright spike of 15 to 30 cm (6–12 in); the spikelets are about 20 mm long, sessile, and arranged on each side of the stem in alternate pairs, occasionally in threes at middle of spike. The leaves are long, broad, rigid, and upright; they have a smooth but furrowed sheath with a very short ligule (a small flap of tissue). It flowers from July to August. The caterpillar of the lyme grass moth (*Arenostola elymi*) feeds on its stems in May and June, as does that of the rustic shoulder knot (*Apamea basilinea*). It is affected by a fungus, the grass-culm smut (*Ustilago hypodytes*). It is locally common on dunes around the coast of the British Isles; rare in Ireland.

Marram Grass (*Ammophila arenaria*)
Family Gramineae (146) **Fig. 9(3)**

Grows only on sandy coasts. The stem is stout and

Fig. 9 1. Sea couch-grass (*Agropyron pungens*). 2. Sand couch-grass (*Agropyron junceiforme*). 3. Marram grass (*Ammophila arenaria*). 4. Sea lyme grass (*Elymus arenarius*). 5. Hastate orache (*Atriplex hastata*). 5a. Bracteole of Babington's orache (*Atriplex glabriuscula*). 5b. Bracteole of *A. hastata*.

erect, between 60 and 120 cm (24–47 in) tall, and rises from a creeping matted root which binds the sand together around its base. Its leaves are long, rigid, and pointed; smooth on the outside, inrolled, and rough within. They are bluish-green with long basal sheaths; the ligule is long, often over 10 mm, pointed, and split into two portions. The flower-head is 7–15 cm ($2\frac{3}{4}$–6 in) long, spike-like, stout, cylindrical, tapering, and dense with spikelets each 12–14 mm long. Marram grass is a perennial and propagated by division; it is very abundant and dominant on some dunes. Flowering period is July and August. The caterpillar of the shore wainscot moth (*Leuconia littoralis*) feeds on it from August to May; and it is affected by a fungus, the grass-culm smut (*Ustilago hypodytes*). Grows around the coasts of the British Isles.

Hastate Orache (*Atriplex hastata*)
Family Chenopodiaceae (34) **Fig. 9 (5 and 5b)**

This dark green, mealy herbaceous annual grows up to 100 cm (39 in), and it has a variable habit, erect or prostrate, and branching. The stems are very often reddish. The lower leaves are triangularly spear-shaped and opposite, with acute and more or less horizontal lobes some of whose margins make an angle of at least 90° and therefore contract very abruptly into the leaf-stem; uppermost leaves are lance-shaped and taper into a short stem; all leaves are entire (not toothed or cut at margin). The tiny green petal-less flowers are clustered on simple or panicled spikes and are leafy below. It flowers July to August and grows commonly both inland, and by the sea on sand and shingle or mud above high water mark.

Babington's orache (*Atriplex glabriuscula*) (Fig. 9, 5a) is very like *A. hastata* and often regarded as just a sub-species. It can only be differentiated with reasonable certainty by the flower bracteoles which are united up to about the middle. It flowers July to August and grows on sandy and gravelly shores around the British Isles.

Common orache (*Atriplex patula*) is very similar to *A. hastata*, the two often growing together; it also has a variable habit but is perhaps more often prostrate than is *A. hastata*. The leaves are possibly the best means of differentiation; these are usually linear to linear-oblong, gradually tapering into a short leaf-stem; only rarely are they triangularly

shaped, and if so the margins make an angle with the leaf-stem quite noticeably less than 90°. Uppermost leaves are stemless. It flowers July to August and grows throughout most of the British Isles on cultivated land or in waste places, also near the sea where it is usually less common than *A. hastata*. Caterpillars of the orache moth (*Trachea atriplicis*) and the dark spinach moth (*Pelurga comitata*) feed on species of orache.

Sea Couch-grass (*Agropyron pungens*)
Family Gramineae (146) **Fig. 9(1)**

A glabrous and often glaucous perennial with a grass habit. It is densely tufted, has erect stems 30–90 cm (12–35 in), and far-creeping underground shoots. The leaves are rough, pungent, rather stiff, and have inrolled margins; their ribs are broad and prominent. Flower-spikes are stout and fairly stiff, some 5–12 cm (2–$4\frac{1}{2}$ in) long, and bear a resemblance to an ear of wheat. Spikelets are closely set, 10–18 mm long, each has three to eight florets; most spikelets overlap the adjoining one by at least half its length (a characteristic). It is in flower from July to September. Grows on dunes where it can be locally dominant, also in saltmarshes and estuaries. Occurs in coastal regions from Cumbria south to Cheshire, along the south coast from Kent to Cornwall, then north to Glamorgan and north-east Yorkshire. Is probably absent in Scotland. In Ireland occurs from Dublin south to Cork, also Limerick and south-east Galway.

Sand Couch-grass (*Agropyron junceiforme*)
Family Gramineae (146) **Fig. 9(2)**

This glaucous perennial has stalks 25–50 mm (10–20 in) tall, and far-creeping, wiry, underground shoots which give rise to the formation of extensive tufts. The stems are bluish-green, stout and smooth; flowering stems are more or less erect; sterile shoots prostrate at first, rising above ground towards tip. Leaves, pointed, stiff, smooth, and flat (inrolled at margin when dry), erect or partially drooping; the ribs are broad and prominent above, densely pubescent with short hairs (a characteristic) thus contributing to the bluish-green appearance. Flower-spikes are 5–15 cm (2–6 in) long, stout, erect and stiff, with the stalkless spikelets arranged alternately on each side of stem; spike-stem is slightly angled at each node thus producing a mildly zig-zagged effect. Flowering period is from June to August. A common species on coastal dunes and sandy coasts throughout the British Isles up to Shetland. The caterpillars of two species of butterfly have couch-grasses amongst their food plants; the Essex skipper (*Thymelicus lineola*), and the ringlet (*Aphantopus hyperantus*).

Common Centaury (*Centaurium erythraea* maritime variety *subcapitatum*)
Family Gentianaceae (100)

This common, erect annual grows up to 50 cm (20 in), and is essentially a maritime species occurring wherever the shore is sandy; but can also be found in dry grassland and the margins of woodland. Usually a solitary stem (sometimes more) rises from the basal rosette, and is branched above. Basal leaves are ovate to oblong, and as a rule are 1–5 cm ($\frac{1}{2}$–2 in) long and 8–20 mm wide. The numerous pale pink flowers are crowded and form a dense cyme; may be stemless or with but a very short one. They are funnel-shaped, the corolla-tube is longer than the calyx and has lobes of 5–6 mm. Stamens are inserted at top of corolla-tube. Very variable. It is in flower from June to October. Occurs commonly on practically all the sandy shores of Great Britain's coastal counties. Its flowers are sometimes visited by the humming bird hawk moth (*Macroglossa stellatarum*).

Slender Centaury (*C. pulchellum*)

This differs from *C. erythraea* in being unbranched, without a basal rosette, and growing 3–15 cm (1–6 in) tall. Its pale pink flowers have stems, do not grow in clusters, and the lobes of corolla are from 3–4 mm. It usually grows in damp places such as saltmarshes; it is locally common near the sea in the coastal counties of southern to central England, far less common to rare northwards to Scotland.

Shore Centaury or Seaside Centaury (*C. littorale*)

This is an erect annual of 2–25 cm ($\frac{3}{4}$–10 in). Its basal leaves are more or less strap-shaped, and form a rosette; they are up to 5 mm broad. The bluish-pink flowers are larger (lobes 6–7 mm) but fewer than those of *C. erythraea*, period July to August.

Seems almost confined locally to the dunes and sandy places along the coasts of Wales, north-west England, and south-west Scotland.

Centaurium capitatum

This is a small erect annual of 2–5 cm ($\frac{3}{4}$–2 in). Basal leaves measure about 20 × 8 mm and are not in a rosette. The flowers are sessile and not clustered; lobes or corolla 5–6 mm, corolla-tube about equal in length to calyx. Stamens are inserted at base of corolla-tube. Usually grows in grass on calcareous soils near the sea. Very local indeed, occurring from Kent to Cornwall, in parts of Wales, Lancashire, and Yorkshire, also Guernsey.

English Stonecrop (*Sedum anglicum*)
Family Grassulaceae (54)

A small, glaucous, succulent little evergreen up to 5 cm (2 in) high. It has numerous short and branching, creeping stems which bear small (3–5 mm), green (occasionally pink-tinged), thickish, ovoid leaves; these clasp the stem alternately and have a tiny spur at the base. Often spreads extensively, forming mats on shingle beaches, cliff banks, and rocks. The inflorescence usually has two main branches which bear from three to six flowers of about 12 mm diameter, each on a short, stout pedicel (stem) of about 1 mm. The somewhat star-like white flowers are tinged pinkish on back. In bloom from May to August. The caterpillar of the northern rustic moth (*Ammogrotis lucernea*) feeds on it from August to May, after the flowering period, and those of the sword-grass moth (*Xylena exsoleta*) feed on it from April to May. It is by no means confined to coastal regions. Perhaps most common on west coasts from the Hebrides southwards; far less so along eastern and southern coasts. Occurs throughout Ireland.

Wall Pepper or Biting Stonecrop (*Sedum acre*)
Family Grassulaceae (54)

Of a similar habit to *S. anglicum* but up to 10 cm (4 in) tall, and its numerous thick, oval, stalkless, succulent green leaves are imbricate (overlap the adjoining leaves); sometimes not imbricate on flower-stems. At the top of each flowering stem is borne a cluster of tiny golden-yellow flowers each about 12 mm in diameter, with lanceolate petals, and star-like. The common name of wall pepper is almost certainly given because of the hot acrid taste produced if the stem or leaves are chewed. Flowering period June to July. Occurs throughout the British Isles but apparently not in Shetland. In coastal regions it is found on dunes, shingle and walls.

Large Evening Primrose (*Oenothera erythrosepala*)
Family Onagraceae (66)

An introduced and usually biennial plant with erect, stiff, and hairy stems of 50–100 cm (20–39 in); hairs on stem have red bulbous bases. Basal leaves broad, lance-shaped, crinkled, and somewhat pubescent, they are distantly toothed, and stalked; stem-leaves are almost stalkless, midrib white or occasionally red. The large pale-yellow flowers grow in spikes, with petals of between 4 and 5 cm (1½–2 in) and broader than they are long. Sepals are striped red

Large evening primrose (*Oenothera erythrosepala*) on a fixed sand dune.

and covered in hairs. As suggested by the common name, its flowers open in the evening. Flowering period is June to September. Near the coast it occurs on dune systems; but may also be encountered inland along roadsides, railway banks, and waste places. In Britain it occurs northwards to Perth and Angus; also Ireland and the Channel Isles.

Small-flowered Evening Primrose (*O. parviflora*)

A biennial to perennial plant which was introduced and is now established on dunes in south-west England. Grows from 10 to 80 cm (4–31 in), stems only sparsely hairy (hairs having red bases). Basal-rosette leaves paddle-shaped; stem leaves narrow and lanceolate, somewhat thick and fleshy, with reddish veins. Flowers are much smaller than in *O. erythrosepala*, the yellow petals 11–18 mm. In flower from June to September.

Fragrant Evening Primrose (*O. stricta*)

This is an annual or biennial with stems between 50 and 90 cm (20–35 in); hairs on stem do not have red bulbous bases. Stem-leaves 3–10 cm (1–4 in) diminishing upwards, broadly lance-shaped with wavy and distantly toothed margins, and almost stalkless. Basal leaves are narrowly lance-shaped. The yellow petals are 3–4 mm (1–1½ in) long, and later turn red. Has a noticeable fragrance. Flowers June to September. An introduced species found in dunes, and now well established in south-west England and the Channel Isles. Occurs in other localities north to Selkirk.

Prickly Saltwort (*Salsola kali*)
Family Chenopodiaeae (34)

This salt-loving, prickly annual grows up to 60 cm (24 in); the stems are spreading or prostrate, only seldom erect; they have pale green or reddish stripes, are furrowed, wavy, and usually much branched. The awl-shaped leaves are 1–4 cm (½–1½ in) long, and although terminating quite bluntly are furnished with a sharp point; they are succulent and sessile. Its small, single pinkish-green flowers are quite inconspicuous, and usually solitary in a leaf axil; each has two leaf-bracteoles. The flowering period is July to August; it grows quite commonly on sandy shores around the coasts of Britain. It has

Prickly saltwort (*Salsola kali*).

Sea holly (*Eryngium maritimum*) on sandy/shingle beach.

been used on a commercial basis to obtain salt. The caterpillars of two moths, the sand dart (*Agrotis ripae*) and the coast dart (*Euxoa cursoria*) feed on it.

Sea Holly (*Eryngium maritimum*)
Family Umbelliferae (75)

One of the most easily recognisable of all the plants on our sandy shores, with a somewhat thistle-like appearance, and intensely glaucous-coloured holly-like leaves. This glabrous, branched perennial grows 30–60 cm (12–24 in). Its leathery, three-lobed radical leaves are rounded, 5–12 cm (2–4½ in) in diameter, and have thickened, spiny, cartilaginous, white margins. The upper leaves clasp the stem and have spiny lobes which emanate from a common centre. The dense, prickly, pale blue or lavender, terminal flower-heads are about 1.5–2.5 cm (½–1 in), and each individual flower is 8 mm in diameter. Below the flower-head is a whorl of stiff, very spinous, leaf-like bracts. Flowering period July and August. The caterpillars of the sand dart moth (*Agrotis ripae*), and of the dingy skipper butterfly (*Erynnis tagas*) feed on it. Occurs on sandy and shingly shores around the coasts of the British Isles north to Shetland.

Sea Kale (*Crambe maritima*)
Family Cruciferae (21)

This perennial maritime herb has an erect and bushy habit, the stems range in size from 40 to 60 cm (16–24 in) by 2–3 cm; its preferred habitat seems to be along the drift line on a shingly beach, occasionally on coastal sands, but rarely found on sea-cliffs. The stem is branched below, and has large, up to 30 cm (12 in), long-stalked leaves; these are thick, glabrous, and bluish-green, with the edges deeply waved, and distantly toothed. Upper leaves are narrow; and the topmost ones very narrow and bract-like with margins entire. The flowers are 10–16 mm in diameter with four white petals 6–9 mm long.

Sea kale (*Crambe maritima*) on a shingle beach at Llanddulas.

They grow in crowded clusters, each flower on a long stalk, and well clear of the foliage. In flower from June to August. The seedpods are up to 14 mm long by 8 mm wide, each containing one seed. When the pods drop off they either germinate around the parent plant or are dispersed by floating in sea water, where they may remain for several days with no adverse effect. Sea kale is by no means a common plant but grows locally along the south coast, also the Isle of Man, and from North Wales to south-west Scotland. It is very rare along the east coast and Ireland.

Sea Rocket (*Cakile maritima*)
Family Cruciferae (21)

This locally common annual has a prostrate or ascending, branched stem of up to 45 cm (18 in); the branches are arranged in a zig-zag fashion. The entire plant has an untidy bushy appearance and often forms large patches along the drift line on sandy or shingly beaches, where it often grows in association with sea kale (*Crambe maritimum*), and sea bindweed (*Calstegia soldanella*), but is equally at home on salt marshes along with sea plantain (*Plantago maritima*), and sea lavender (*Limonia vulgare*). Its fleshy, bluish-green lower leaves are 3–6 cm (1–2¼ in) long, and deeply cut into several narrow blunt lobes; they taper into a stalk-like base. Upper leaves are sessile, may be lobed or entire, and without a stalk. Its lilac, purple or white flowers are in terminal clusters on the main stem and branches; the petals are 6–10 mm, twice as long as the sepals. The flowering period is June to August; when the pods drop they either germinate close to parent plant or are dispersed by floating in the sea. The caterpillars of the sand dart moth (*Agrotis ripae*) feed on the plant. Sea rocket occurs in suitable places all round the British Isles.

Sea Sandwort (*Honkenya peploides*)
Family Caryophyllaceae (30)

Some of the older reference books refer to this plant as sea purslane. Sea sandwort is a succulent, salt-loving perennial with a creeping habit, and often covers sizeable areas of sandy shingle shores. Its prostrate stems produce forked flowering-shoots of up to 25 cm (10 in); these have very fleshy, stemless, broadly oval, dark green leaves some 6–8 mm long

Sea sandwort (*Honkenya peploides*) growing on fine shingle at Slapton Sands.

Sea rocket (*Cakile maritima*).

which are arranged in opposite rows along the shoots. The greenish-white flowers are five-petalled, and about 6–10 mm in diameter; they are solitary in the axils of upper leaves and in the stem forks. Petals of female flowers are shorter than the sepals, but in male flowers petals and sepals are of equal length. Sepals ovate, blunt and fleshy. Flowering period is May to August. A common plant on mobile sand and sandy shingle all round the British Isles. The caterpillar of the bordered straw moth (*Heliothis peltiger*) feeds on it.

Sea Spurge (*Euphorbia paralias*)
Family Euphorbiaceae (78)

A locally common bushy perennial of 20–40 cm (8–16 in) with an erect or ascending habit. It produces several fertile and sterile stems. The very leathery leaves are alternate, numerous, 5–20 mm long, glaucous, thick, ovate or oblong, obtuse to subacute, and often overlapping; they grow in whorls up the stem. Its yellowish-green flowers grow in umbels, each comprised of three to six rays. The seeds are smooth with a very small caruncle. Flowering period July to October. Occurs locally on sandy coasts southwards from Wigtown and Norfolk, also Channel Isles, and coasts of Ireland where it is rare in the north and west. The caterpillar of the spurge hawk moth (*Celerio euphorbiae*) feeds on it during August and September.

Portland Spurge (*E. portlandica*) is usually smaller than *E. paralias* at 5–40 cm (2–16 in), and has less leathery leaves which are often narrower. The seeds are pitted, not smooth. Usually flowers a little earlier, May to August. It grows very locally in sandy coastal habitats westwards along the south coast from Sussex, and up the west coast to Wigtown; also the Channel Isles, and the coasts of Ireland where it is rare in the north-west.

The now very rare, and almost extinct in Britain, purple spurge (*E. peplis*) is a procumbent maritime annual. It is usually four-branched, each 1–6 cm ($\frac{1}{2}$–$2\frac{1}{4}$ in) long, often purplish, but may be glaucous. As a rule the leaves are much smaller than those of *E. paralias* and *E. portlandica*, 3–10 mm, and have a large rounded ear-like lobe on one side of base. The seeds are smooth and not caruncled. Now it occurs only very sporadically in its former shingle beach sites, which are principally in Devon, Cornwall, and Channel Isles. Flowers July to August.

Sea Bindweed (*Calystegia soldanella*)
Family Convolvulaceae (104)

This seaside species of bindweed has long, prostrate, smooth, and reedish creeping stems of 10–60 cm (4–24 in) and is unlike other bindweeds whose stems have a climbing habit, this being possible because of their propensity to twist in a counter-clockwise direction. The leaves are fleshy, dark green, and kidney-shaped; leaf-blade 1–4 cm ($\frac{1}{2}$–$1\frac{1}{2}$ in) long, leaf-stem usually a little longer than blade. When fully unfurled the solitary flowers are funnel-shaped, up to 5 cm (2 in) long, and the corolla 2.5–4 cm (1–$1\frac{1}{2}$ in) across; they are pink or mauve and have five narrow interior white stripes. Bracteoles are 10–15 mm long, more or less oblong with a rounded tip, and shorter than the calyx. The flowering period of this perennial is June to August. It grows locally on sand dunes and sandy shingle, mainly along the coasts of England and Wales. Much less common in Scotland and Ireland, where it becomes rare along the northern coasts.

Sea Stork's-bill (*Erodium maritimum*)
Family Geraniaceae (39)

The habit of this small, locally common, annual is prostrate with decumbent and slightly hairy stems of up to 15 or 20 cm (6–8 in). Its leaves are simple, and oval to heart-shaped with coarsely-toothed lobes; leaf blade 5–15 mm; leaf-stalk relatively long. The flowers are borne singly or in twos on axillary peduncles, and with or without small pink petals which, if present, are not longer than the calyx. Sepals are about 4 mm long, oblong, and hairy with a short terminal point. Flowering period is May to September, the fruit is up to 1 cm ($\frac{1}{2}$ in) long and beak-like, hence the common name. It likes an open habitat such as fixed dunes, and dry stony or grassy areas close to the sea. A local species in suitable coastal areas from Kent to Cornwall, and along the west coast north to Wigtown. Also occurs in Norfolk, Durham, Northumberland and the Channel Isles. In Ireland southwards from Down on the east coast, and up the west coast to Clare.

Bloody Crane's-bill (*Geranium sanguineum maritime* var. *prostratum*)
Family Geraniaceae (39)

The procumbent (not erect or ascending) stems of

Sea bindweed (*Calystegia soldanella*) on sand dunes.

this locally common perennial are 10–40 cm (4–16 in) long; they are numerous, slender, hairy, and abruptly bent at the nodes. They bear roundish leaf-blades 2–6 cm (1¾–2¼ in) in diameter which are deeply cut into five or seven lobes; each lobe is itself further divided into three segments but not to the base, and the secondary lobes themselves may be similarly further divided. Leaf lobes of the maritime variety *prostratum* are often broader than those of the form found inland. The large flowers are about 4 cm (1½ in) across with bright purplish-crimson petals which are shallowly notched and about 12–18 mm long; they are borne singly on long stalks. Flowering period is July to August; seed pods about 3 cm (1¼ in) long and pointed. The maritime variety grows on fixed dunes and other sandy places, occa-sionally on sea-cliffs. May be found locally in suitable places around most of the British Isles with the possible exceptions of south-east England and southern Ireland.

Tamarisk (*Tamarix anglica*)
Family Tamaricaceae (27)

This attractive shrub has been introduced in many coastal areas in the southern half of Britain where it is now naturalised. Along with sea buckthorn (*Hippophae rhamnoides*), marram grass (*Ammophila arenaria*), sand sedge (*Carex arenaria*), and others, it has served to help bind together the would-be shifting sands along our south-east coast. It grows up to 3 m (10 ft) tall, so also helps as a windbreak. The delicate reddish or purplish twigs bear narrow scale-like leaves of up to 2 mm long, they are pointed, pale green, and overlapping. Tapering plumes, 1–3 cm ($\frac{1}{2}$–$1\frac{1}{4}$ in) long, of tiny five-petalled pink or white flowers each only 3 mm long, are borne terminally along the current year's growth, thus giving the tamarisk an overall feathery appearance. The flowering period is July to September. It occurs along the east coast from Sussex southwards, and then westwards to Cornwall and the Channel Isles. Another species, *T. gallica*, which has a more slender inflorescence, has also been introduced but is far less common. It occurs within the same range.

Viper's Bugloss (*Echium vulgare*)
Family Boraginaceae (103)

A strikingly beautiful but very rough, erect, biennial which grows from 30–90 cm (12–35 in) tall. The first year it produces just a rosette of narrow, dull green leaves; these are very hairy and up to 15 cm (6 in) long. During the second year it develops into a tall bushy plant with thick bristly stems, the upper half of which gives rise to many short leafy branches, each bearing from 8 to 12 bright blue, funnel-shaped, tubular flowers with protruding stamens. The flowers, each 15–18 mm and subsessile, are arranged in a terminal panicle, but only three or four are open together; when in bud they are pinkish-purple. Viper's bugloss is locally common on light dry soils, gravelly dunes and sea-cliffs; is also fond of calcareous soils. It has a scattered distribution throughout England and Wales; less common is Scotland and Wales. It flowers from June to September. It is known to be one of the plants fed upon by caterpillars of the orange swift moth (*Hepialus sylvina*) (which feed on its roots), also the marbled clover moth (*Heliothis dipsaceus*), the small angle shades moth (*Euplexia lucipara*), and the painted lady butterfly (*Vanessa cardui*) (which eat the foliage). The viper's bugloss moth (*Anepia irregularis*) often rests on the plant, hence its name, but the caterpillars do not feed on it.

Yellow Horned-poppy (*Glaucium flavum*)
Family Papaveraceae (19)

This free-flowering perennial or annual herb has an erect branched stem of 30–90 cm (12–35 in), and is very easily identified. It requires a saline soil and is locally common on coastal shingle in scattered areas around our shores. The entire plant is glaucous green and has a succulent texture so unusual amongst poppies. Its stems are sturdy and void of hairs. The basal leaves are stalked, pinnately lobed, and covered with rough hairs; the wavy lobes are themselves coarsely toothed. Upper leaves are less deeply lobed, rough, and partly clasp the stem. The flowers are 6–9 cm ($2\frac{1}{4}$–$3\frac{1}{2}$ in) in diameter and have four roundish yellow petals which are crinkled until fully opened. The seed pods are curved and measure up to 30 cm (12 in), their resemblance to 'horns' is typical of the genus and obviously gives rise to the common name of 'horned-poppy'. Flowering period is June to September. It has a coastal distribution south from Kincardine and Argyll, including Ireland and the Channel Isles. A point worth remembering is that the yellow juice which exudes from a cut stem is poisonous.

Sea Radish (*Raphanus maritimus*)
Family Cruciferae (21)

This somewhat uncommon, erect perennial or biennial, has stems 20–80 cm (8–31 in) long and bristly, especially the upperparts. Lower leaves are coarse, large, dark green, pinnately divided, with a large terminal lobe and from four to eight pairs of contiguous lateral lobes which are smaller and often alternating in size; the very small basal lobes point downwards. Upper leaves are smaller and also dark green. Flowers are veined, on long stalks, and arranged in a loose sort of flowerhead. Petals about 2 cm ($\frac{3}{4}$ in) and yellow, rarely white except in Channel Isles. Flowering period June to August. One to five seeds; beak slender, 5–8 mm in diameter, not more than twice the length of top joint, and deeply constricted between the seeds. Does not easily break at joints into one-seeded segments as does the wild radish (*R. raphanistrum*) whose beak is up to five times as long as the top joint. The seeds of the sea radish are dispersed by floating in sea water. A plant of the drift line along sandy and rocky shores, it also grows on sea cliffs. It occurs locally in south-west England, but rare elsewhere except the Channel

Isles where it is quite common and often white-flowered.

Scentless Mayweed (*Tripleurospermum maritimum* maritime ssp. *maritimum*)
Family Compositae (120)

This perennial, maritime subspecies has a prostrate or decumbent habit with widely spreading stems of 10–30 cm (4–12 in). Leaves have a somewhat oblong outline and may be two or three times pinnate; with ultimate segments short, fleshy, blunt, and more or less tube-like. The long-stalked, solitary flowerheads are terminal and 3–4.5 cm ($1\frac{1}{4}$–$1\frac{3}{4}$ in) across, they have between 20 and 30 spreading, white ray-florets, and yellow disk-florets. In flower July to September, and as the name suggests are without scent or have but little. It is locally common along driftline between shingle beaches and dunes, also sea-cliffs and rocks. Occurs from north Wales and northern England to Scotland, also Ireland. The sub-species *inodorum* has a maritime variety *salinum* which as a rule is a biennial. This has more or less erect stems of 15–60 cm (6–24 in); these may be simple or form a corymbe above. The leaves are similarly erect with short, fleshy, crowded, linear segments which terminate in a short narrow point. Flowerheads are white, 1.5–3.5 cm ($\frac{1}{2}$–$1\frac{1}{3}$ in) across with 12–22 ray-florets, and yellow disk-florets. Flowering period and habitat as for *maritimum*, but distribution seems somewhat confined to south of England and South Wales.

Creeping Willow (*Salix repens* ssp. *argentea*)
Family Salicaceae (89)

A dwarf deciduous shrub, 30–150 cm (12–59 in), with slender, prostrate to erect stems; it is branched, bushy, of a creeping habit, and quite easily identified. When young, the twigs have a silky pubescence, and are finely striated under bark; the young oval buds are similarly pubescent but soon become glabrous. The leaves of first-year sterile shoots are up to 4.5 by 2.5 cm ($1\frac{3}{4}$ by 1 in), usually rounded at base, or occasionally somewhat heart-shaped basally; those of fertile shoots are considerably smaller and, as a rule, silky on both sides. All leaves are alternate and simple. The catkins appear before the leaves and are either subsessile or borne on short leafy stalks; the sexes are on different plants. Male catkins are 5–20 mm, slender and ovoid to oblong. Those of the female are a little larger, 8–25 mm, and globose to oblong. However, the catkins of both sexes tend to be bigger than those of the ssp. *repens*. Flowering period April to May. Distribution covers all but a few coastal counties in the British Isles. Occurs, and is sometimes dominant, on dune slacks, and occasionally grows on rocky heathland in northern Scotland. The flowers (catkins) are visited by the honey-bee (*Bombus terrestris*), and the plant is subject to attack by the fungus *Melampsora repentis*. The caterpillar of the small chocolate-tip moth (*Clostera pigra*) feeds on it from June to September.

Sea Buckthorn (*Hippophaë rhamnoides*)
Family Elaeagnaceae (65)

This thorny, thicket-forming, deciduous shrub is also known as sallow buckthorn. It grows up to 3.5 m ($11\frac{1}{2}$ ft) high and is freely suckering. The leaves have short petioles, and unfold during April, they are linear with margins entire, and up to 8 cm (3 in) long after flowering; upper surface is dull green with a scattering of silvery stellate scales; underside silvery-grey and scurfy. Flowering period is March to April either before or with the leaves. Flowers are very small, inconspicuous, and greenish; female flowers grow in crowded clusters, and the male flowers in small spikes in axils of lowest bracts. The fruit is quite unmistakable and very conspicuous, the orange-coloured, berry-like drupes are borne in clusters and often persist through the winter months; each drupe is about 8 mm in diameter, and much sought-after by wintering birds, especially members of the thrush family such as redwings and fieldfares. The macro moth (*Gelechia hippophaella*) larvae in spun shoots and the leaf-hopper (*Psylla hippophaes*) cause leaf distortion. Sea buckthorn grows on fixed dunes and occasionally sea-cliffs. Occurs as a native on parts of the east coast from Yorkshire down to Sussex where it is dominant in suitable areas. Elsewhere it has probably been planted, and is now dominant on parts of the Lancashire coast and along some of the sandy bays of north-east Scotland. Occurs locally in many other parts of the British Isles.

Land Snails and Slugs Found near the Coast

See Colour Plate.

Vertigo pulsilla
Family Vertiginidae; Sub-family Vertigininae **(10)**
2×1.1 mm. Shell pale yellowish brown and glossy; distinctly conical. Occasionally in sand dunes but mainly inland.

Vertigo pygmaea
Family Vertiginidae; Sub-family Vertigininae **(12)**
$1.7–2.2 \times 1–1.2$ mm. Shell pale to dark brown and usually dullish. Occasionally in sand dunes but mainly inland.

Vallonia pulchella
Family Vallonidae; Sub-family Valloniinae **(8)**
2–2.5 mm. Pale translucent shell; rather glossy. Occasionally in sand dunes but mainly inland.

Arion subfuscus
Family Arionidae **(11)**
A medium slug 5–7 cm ($2–2\frac{3}{4}$ in). Usually dark brown with a darker longitudinal band on either side. Occasionally in sand dunes but mainly inland.

Vitrina pellucida
Family Vitrinidae **(6)**
4.5–6 mm. Shell usually pale green; somewhat globular; smooth, glossy, translucent and very thin. Often very abundant in the grassy hollows of coastal sand dunes.

Milax gagates
Family Milacidae **(7)**
A medium slug 5–6 cm ($2–2\frac{1}{4}$ in). Black or dark grey; has 14 paler longitudinal grooves on either side. Often in grassy places near the sea.

Deroceras caruanae
Family Limacidae **(5)**
A small to medium slug 2.5–3.5 cm ($1–1\frac{1}{2}$ in). Light to medium brown, but occasionally greyish to black. Often in hedges and fields in some areas near the sea.

Candidula intersecta
Family Helicidae; Sub-family Helicellinae **(1)**
$5–8 \times 7–13$ mm. Shell white to ginger, often darkly banded spirally; opaque; globular and depressed. Especially in dunes but mainly inland.

Cernuella virgata
Family Helicidae; Sub-family Helicellinae **(9)**
$6–19 \times 8–25$ mm. Shell white to ginger, usually darkly banded spirally; globular with a high convex spire. Especially in dunes but mainly inland.

1. *Candidula intersecta*. 2. *Helix aspera*. 3. *Cochlicella acuta*. 4. *Ponentia subvirescens*. 5. *Deroceras caruanae*. 6. *Vitrina pellucida*. 7. *Milax gigates*. 8. *Vallonia pulchella*. 9. *Cernuella virgata*. 10. *Vertigo pulsilla*. 11. *Arion subfuscus*. 12. *Vertigo pygmaea*. 13. *Theba pisana*.

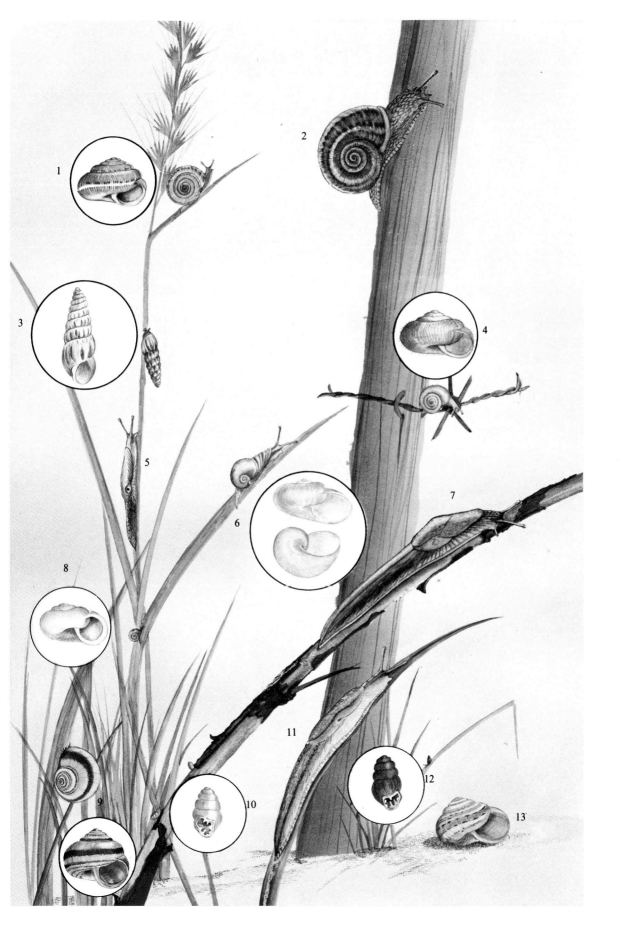

Cochlicella acuta

Family Helicidae; Sub-family Helicellinae (3)

10–20 × 4–7 mm. Shell a very elongated cone; white to ginger with darker bands; very variable. A maritime species, usually in dunes and coastal grasslands.

Ponentia subvirescens

Family Helicidae; Sub-family Hygromiinae (4)

4–6.5 × 5–8 mm. Shell globular and slightly depressed; dull greenish yellow; somewhat translucent; covered with short hairs. Grassland and rocky areas, usually damp and close to sea.

Theba pisana

Family Helicidae; Sub-family Helicinae (13)

9–20 × 12–25 mm. Shell globular and slightly depressed; white to ginger; varied pattern of dark spiral bands. Usually near the sea and almost exclusively so on dunes at the north of its range.

Helix aspera

Family Helicidae; Sub-family Helicinae (2)

25–35 × 25–40 mm. Shell usually pale brown, sometimes yellow; has up to five dark spiral bands. Its wrinkle-like sculpturing is characteristic. Very varied species. Found mainly inland but also in dunes.

Sites of Specific Interest

Initial figures refer to locations on Map 2 on page 132.

1. BRAUNTON BURROWS (NORTH DEVON)

Braunton Burrows is one of the most extensive areas of sand dunes in Britain, extending northwards from the estuaries of the rivers Taw and Torridge. About two-thirds of the area have been designated a National Nature Reserve. The whole of the dune system is owned by the Christie Estate Trustees, although most of it is leased to the Ministry of Defence and used for military training. It is this section that has been sub-leased to the Nature Conservancy Council since 1964 and is now managed by it. Military activities take priority; access to certain parts is not always possible.

Braunton Burrows is best known for its flora and has attracted botanists since the seventeenth century. Over 400 species of flowering plants have been recorded on the reserve, which is composed of a mixture of high dunes, damp slacks and hollows. On the fixed dunes, such species as viper's bugloss,

Visitors to Braunton Burrows National Nature Reserve are clearly warned of its occasional use as a Ministry of Defence Training Ground.

Map 2   Shingle beaches, sandy shores and sand dunes.

evening primrose and biting stonecrop are common in summer. In the damper slacks, marsh orchids grow in profusion together with marsh helleborines and the round-leaved wintergreen. Like other dune systems across the country, sea buckthorn is a problem and will soon swamp large areas if allowed to spread. However, careful management of the reserve has resulted in the removal of large sections of buckthorn thickets in recent years.

Rabbits are by far the most common animal seen on the reserve, although numbers have declined dramatically since the myxomatosis outbreaks of the last thirty years. Other mammals found on the reserve include fox, hedgehog, weasel, mole, mink and the smaller voles, shrews and mice.

Since the area is on the west coast migration route, passage birds use the reserve to rest and feed. In summer, a wide variety of butterflies are attracted to the rich flora, especially the common blue, meadow brown, gatekeeper and the handsome dark green fritillary. Of the day-flying moths, the burnet moth is most noticeable.

The reserve has a full-time warden and a number of voluntary assistants. Guided tours can be arranged for groups and school parties, and further information is obtainable from the warden at Broadeford Farm, Heddon Mill, Braunton, North Devon (Tel: 0271 812552). Two free car parks are provided for visitors. The one at the southern end of Sandy Lane is best approached by turning off the B3231 Braunton to Saunton road. The other is at Broadsands.

2. SLAPTON SANDS (DEVON)

Slapton Sands is an excellent example of a shingle bar and forms part of a 184-hectare (460-acre) nature reserve managed by the Field Studies Council. The long freshwater lake behind the shingle ridge is known as Slapton Ley, which is also the name given to the Council's study centre in the nearby village of Slapton.

The ridge has a main road along it and is therefore subjected to considerable public pressure from visitors, especially during the holiday season. Nevertheless, shingle flora abound throughout the summer with superb growths of yellow horned-poppy at the northern end and both thrift and campion lining the roadway.

Slapton Sands.

The Ley has been an informal sanctuary for birds since 1896. Most species of tern pass through during spring and summer migration. The northerly part, known as Higher Ley, has a sizeable area of reed marsh in which Cetti's warbler breeds in summer. A recent arrival, there are fewer than 200 pairs nationwide. The Lower Ley supports numerous wildfowl in winter and great-crested grebe breed there annually.

The ridge is reached by taking the A379 Kingsbridge to Torcross road. Car parking is available at intervals along the whole ridge. More information is available from the Field Studies Centre, Slapton, Devon.

3. CHESIL BEACH (DORSET)

Chesil Beach is a classic shingle bar and is one of the longest and finest shingle ridges in Europe. Plants grow only on the stable beach crest and on the landward side of the bar. They include typical shingle species.

The bar encloses a freshwater lagoon known as the Fleet, which has extensive reedbeds and is an important habitat for both migrant and resident birds. Common and little terns nest along the ridge, so that the reserve demands careful management during the summer when visitor disturbance is high. This is successfully carried out by members of the Dorset Trust for Nature Conservation.

The western end of the shingle ridge can be reached by turning left off the B3157 Weymouth to Bridport road at Abbotsbury. Parking on beach.

4. WEST BEXINGTON (DORSET)

Further west from Chesil Beach exists an 18-hectare (45-acre) reserve managed by the Dorset Trust for Nature Conservation. The shingle flora in the area is typical, but behind the beach is a reedbed and scrub-covered slope of great importance to both breeding and migrant birds. Access to the marshy area is not permitted but there are no restrictions along the beach.

The reserve is approached by turning off the B3157 at Swyre towards West Bexington. Park on the beach.

5. STUDLAND HEATH (DORSET)

This National Nature Reserve of over 160 hectares (400 acres) lies to the north of Studland village and comprises sand dunes, reed beds, heathland, woodland and a freshwater lake.

The lake provides a resting place for wintering wildfowl, including pintail, wigeon, pochard and tufted duck. Reed warbler, reed bunting, Dartford warbler and water rail also occur. Otter, fox and roe deer are present. The adder, grass snake and the rare smooth snake are found on the reserve, as are the sand lizard, common lizard and slow worm. Butterflies, dragonflies (over 20 species) are numerous and the old hazel wood supports many birds.

Plants found on the reserve include bog bean, reedmace, small quillwort, yellow bog asphodel, marsh gentian, sundew, pale butterwort, bladderwort, Dorset heath and sea bindweed.

A car park lies at the southern end of the reserve,

near the Knoll House Hotel. Two nature trails have been set up – the sand dune trail operates throughout the year and the woodland trail from April to September. An observation hut opens on Sundays. Further information is available from the Warden, 'Coronella', 33 Priest's Road, Swanage, Dorset BH19 2RQ (Tel: Swanage 423453).

6. BROWNSEA ISLAND (DORSET)

Owned by the National Trust, 80 hectares (200 acres) of this 200-hectare (500-acre) island are leased to the Dorset Trust for Nature Conservation as a nature reserve, offering a range of habitats including saltmarsh, sandy-shore, heathland and woodland. Both sandwich and common terns nest on the reserve, together with a colony of black-headed gulls and the more typical oystercatcher. A large heronry exists on the island and there is a good chance of seeing the red squirrel in one of its few remaining southern haunts.

The island is reached by passenger ferry from Sandbanks or Poole Quay and is open every day from Easter to October. Guided walks are arranged for summer afternoons.

7. DUNGENESS (KENT) RSPB RESERVE

Dungeness is the largest shingle structure in Europe and seems to have originated from a simple spit running across a wide bay now occupied by Romney Marsh. Some 480 hectares (1,200 acres) or so are now owned and managed as a nature reserve by the RSPB.

The reserve is best approached along the Lydd/Dungeness road. A gravel track by Boulderwall Farm leads to the car park and information centre. The site is open throughout the year, but only on certain days of the week, and in winter may only be open at weekends.

Since the area is exposed to the prevailing 'south-westerlies', vegetation is sparse. Broom and bramble predominate, although there are a few large clumps of gorse, which provide nesting places for several of the smaller birds during the summer, including linnets, whitethroats and yellowhammers.

A surprising variety of flowering plants grow on the shingle, although many are in stunted forms because of the exposed nature of the site. Viper's bugloss, ragwort, woody nightshade and wood sage are abundant. It is also a site for the rather local Nottingham catchfly and rare stinking hawksbeard. The damper areas provide an excellent variety of reeds and sedges where reed buntings, reed warblers and sedge warblers nest.

In all, some 270 species of birds have been recorded, and waterfowl in particular are attracted to the various freshwater pits on the reserve, including such species as goldeneye, goosander, teal, shoveler and pochard. The water areas also attract large numbers of migrant waders, such as green, wood and common sandpipers.

Dungeness is also well-known for its migrant insects, and many entomologists operate mercury vapour lamps in the area during the summer nights, attracting such moth species as pygmy footman, Sussex emerald and several of the hawk-moths. By day, the burnets and cinnabars are seen regularly with the yellow and black caterpillars of the former being very common on ragwort.

Fewer breeding butterflies occur, the most interesting being the Essex skipper. However, the common blue, small copper and gatekeeper are plentiful. Huge clouds of migrants frequently turn up from the sea, mainly large and small whites, but occasionally clouded yellows, red admirals, peacocks and painted ladies pay a visit.

Resident mammals include rabbit, hare, fox, stoat and weasel, together with the smaller field vole and wood mouse. Of the amphibians, newts and toads inhabit the damper places and marsh frogs have successfully spread onto the reserve after their introduction to a pond on Romney Marsh in 1935. On warm, summer days, grass snakes and common lizards are often seen in sheltered places.

A visitor route of 2.5 km (1½ miles) is provided with three hides available from which to observe key places. Visitors are asked to report to the Reception Centre on arrival between 10.30 am and 5 pm Non-members of the society are charged a small fee. More detailed information is available from the Warden, Boulderwall Farm, Dungeness Road, Lydd TN29 9PN.

8. SANDWICH BAY (KENT)

Sandwich Bay Nature Reserve is owned by the Na-

tional Trust, the Royal Society for the Protection of Birds and the Kent Trust for Nature Conservation.

The most interesting feature of the reserve is a sand and shingle spit known as Shell Ness. It is still progressing north-eastwards and its gradual development can be traced back to maps several centuries ago. As a result of this development, saltings have been formed on the landward side, behind the dune ridge. The richest saltmarsh is just behind and to the south-west of Shell Ness. Several species of glasswort grow there, as well as sea lavender, sea aster, annual seablite and sea purslane. Nearer the low-water mark the saltings are devoid of vegetation, apart from the occasional clump of cord grass. The invertebrate life is rich, however, providing valuable food for wading birds.

Sandwich is close to the Continent and is therefore a landfall for migrants. Spring and autumn brings in a variety of species to the reserve. Most feed and rest before leaving for other places. Migrant butterflies like the clouded yellow and painted lady also arrive throughout summer and early autumn.

All birds are recorded by the Sandwich Bay Bird Observatory. The Observatory publishes a full annual report and also provides hostel accommodation for visitors to the area. Details are available from the Secretary, SBBO, Old Downs Farm, Sandwich.

The sand dunes at Sandwich are still forming and the marram grass and sea couch grass help to stabilise the fragile system. Other plants include sand spurge, sea holly and sea convolvulus.

Access by car to the reserve is restricted to no closer than about 2.5 km (1½ miles) from the reserve. Two routes are available and perhaps the easiest is via the toll-gate (fee payable) to a car park just south of Prince's Golf Clubhouse. From here, walk along the shore or track inside the dunes to the south-east corner of the reserve.

Further details are available from the Kent Trust for Nature Conservation, 125 High Street, Rainham, Kent.

9. COLNE POINT (ESSEX)

This Essex Naturalists' Trust reserve is composed of two shingle spits and an extensive saltmarsh at the mouth of the River Colne. Typical plant species of shingle habitats grow on the reserve, with a few cast-coast rarities such as sea holly, sea bindweed and yellow-horned poppy. Waders are numerous during winter and brent geese also occur in large numbers. The reserve has a small colony of nesting little terns.

The reserve is reached from the B1027 road through St Osyth. Care must be taken during very high tides, however, as the car parking space near the sea wall is liable to flood. A booklet, 'The Birds of Colne Point', is available from the Essex Naturalists' Trust Office at South Green Road, Fingringhoe, Colchester C05 7DN.

10. LANDGUARD (SUFFOLK)

The reserve comprises the southern half of the former Landguard Common. Some 15.5 hectares (39 acres) in all, it is mainly composed of shingle coastline and the flora is typical of this habitat, with sea kale, yellow horned-poppy, sea bindweed, sea holly and the locally common sea pea prominent.

A small colony of little terns nests on the southern end of the foreshore during some years. Other breeding birds include ringed plover and black redstart.

Access to the reserve is via Langer and Carr Roads, Felixstowe, using the left turn signposted 'Dock Viewing Area'. The site is towards Landguard Point and is situated between the old fort and the sea. Further details are available from the Suffolk Trust Offices, St Edmund House, Ropewalk, Ipswich IP4 1LZ.

11. HAVERGATE ISLAND (SUFFOLK) RSPB RESERVE

This island of 108 hectares (270 acres) in the River Ore is surrounded by saltmarsh and shingle beaches. Artificially-maintained lagoons support Britain's main breeding colony of avocets. Other bird species include redshank, ringed plover, sandwich tern, teal, wigeon, gadwall and shoveler.

Over 130 plant species have been recorded and include sea purslane, sea lavender, English stonecrop and yellow vetch. Seventeen species of butterfly have been recorded, including the Essex skipper. 'Röesel's' bush cricket can often be heard in summer. Hares are plentiful and the alien coypus have

to be controlled because of damage to river banks.

To reach the island it is necessary to take a boat from Orford Quay. The Quay is approached by following the A12 from Tunstall and then taking the B1084. Visiting permits are by written application only, to the Warden, 30 Mundays Lane, Orford, Woodbridge IP12 2LX. There is a boat-charge for both members and non-members. All visits are escorted. The reserve is open on certain days of the week and details of times of the boat trips and escorted tours are available from the warden.

Car parking for the island is available at the back of Orford Quay. The island itself maintains a Reception Centre, picnic area, toilets and various hides.

12. MINSMERE (SUFFOLK) RSPB RESERVE

The reserve covers some 600 hectares (1,500 acres) of reedbeds, heath, deciduous woodland and shallow lagoons with islands. It provides a breeding place for over 100 bird species, which accounts for the reserve's international importance. Little terns nest on the beach; avocet and common terns in the artificial lagoons; bittern, water rail, marsh harrier,

Looking out over Minsmere.

sedge and reed warbler in the reedbeds; redstart, nightjar and woodcock in the woodland; many more waders and wildfowl winter here.

The reserve is approached either from the B1122 road between Leiston and Yoxford, east of East Bridge, or from the B1125 Dunwich/Minsmere road from Westleton Village. The National Trust provides an all-year car park at Dunwich cliffs. However, disabled visitors are allowed to park near the RSPB reception centre, but should contact the warden prior to doing so.

There is a shop and picnic area at the reserve. Several nature trails and hides with disabled facilities exist and a large public hide is located on the shore, south of the National Trust car park. Non-members of the RSPB are charged for visiting the reserve and all visitors must report to the reception centre on arrival to obtain the necessary permit.

The reserve is open throughout the year, but only on certain days of the week. Visiting dates and further information available from the Warden, Minsmere Reserve, Westleton, Saxmundham, Suffolk IP17 3BY (Telephone: Westleton 281).

13. BLAKENEY POINT (NORFOLK)

Blakeney Point is at the end of a shingle spit some 5 km (3 miles) long and originating from the coast at Cley. It is open to the public free of charge and throughout the year. It can be reached from the beach at Cley by walking westwards, but for the less active an easier approach is by boat from Morston or Blakeney Quays.

Dunes have formed on parts of the shingle spit. These have stabilised recently due to the increased vegetation. The Point has been a nature reserve since 1912 and arouses much interest amongst naturalists and geographers alike. It is one of Britain's most renowned sites for nesting terns and other shore-nesting birds. It is also a superb place to observe migration, with many rarities having been recorded over the years. During the winter, several species of duck and wader haunt the harbour area, as well as large numbers of brent geese.

Boats land at Pinchen's Creek and visitors are strongly advised to take sensible footwear and clothing. A warden and his assistants are readily available to answer any questions, and leaders of large parties are expected to make prior arrangements with the warden.

Facilities for the disabled are good and include a 120-m (400-foot) walkway linking the Lifeboat House to the main observation hut. Inside the Lifeboat House there exists an excellent display describing how the spit evolved and the flora and fauna to be found there. A small shop sells guidebooks, postcards and refreshments.

Yellow horned-poppy (*Glaucium flavum*) along a shingle ridge at Blakeney point.

14. HOLKHAM NATIONAL NATURE RESERVE (NORFOLK)

This is an extensive reserve consisting of some 160 hectares (400 acres) of coastal marshes and dunes, together with 2,200 hectares (5,500 acres) of intertidal sand and mudflats. The area, which is partly owned by the Holkham Estate and Crown Estate Commissioners, was declared a National Nature Reserve in 1967. It is now managed by the Nature Conservancy Council in co-operation with the Holkham Estate, and is the largest coastal national nature reserve in England.

Access to the dunes and beach is unrestricted, but visitors are asked to keep to the paths and off farmland. The western part may be approached on foot along the sea wall from Overy Staithe and from the beach at Wells. Otherwise, Lord Leicester allows access by vehicle along a private road known as Lady Ann's Road, from Holkham village to Holkham Gap. There is a footpath running along the southern end of the saltmarsh at the eastern end of the reserve between Wells and Stiffkey.

Wells, in fact, almost divides the reserve in two. The extensive saltmarsh area is to the east and supports a large number of wildfowl, especially in winter. To the west, what was once a large area of saltmarsh has been reclaimed into marshland pasture over the years. Part of the sand dune system on the seaward side is covered with Corsican pine. Further west, it is not afforested and is much more interesting as far as flora and fauna are concerned.

15. SCOLT HEAD NATIONAL NATURE RESERVE (NORFOLK)

Scolt Head Island is jointly owned by the National Trust and Norfolk Naturalists' Trust, but in 1953 it was leased to the Nature Conservancy Council and in fact present management is by a committee made up of representatives of all three organisations.

The island is not only well-known for the large colony of sandwich terns breeding there, but also for the variety of other plant and animal life. Autumn and spring are excellent times to observe migrating birds and, as with the rest of the North Norfolk coast, birdwatchers flock in their thousands at these times of the year.

Summer trips across to the island are available from Brancaster Staithe. Visitors are strongly advised not to walk across the marshes to the island. During the period early May to late July, entry is not allowed to the ternery at the western end. At this time of year, too, dogs are not allowed anywhere on the island. Even at other times of the year they must be kept fully under control.

16. HOLME BIRD OBSERVATORY (NORFOLK)

This 2.5-hectare (6-acre) reserve, managed by the Norfolk Ornithologists' Association, is a key migration point with close on 300 recorded species to date. Visitors are welcomed at the reception centre, where a full-time warden does the daily recording of any migration taking place. There are four observation hides open throughout the year.

The reserve is about 1 km (half a mile) to the north of Holme village and may be approached on foot along the Thornham sea wall. Alternatively, access by car is down Holme Beach Road to just before the golf course, and then turning right.

Permits are allocated on arrival, but visitors must appreciate that the reserve is only open to the public during the period 10.30 am to 4 pm.

17. HOLME MARSH BIRD RESERVE (NORFOLK)

The Norfolk Ornithologists' Association manage this 35.5-hectare (89-acre) reserve jointly with the Courtyard Farm Trust. It consists of a rough grazing marsh which in winter supports large numbers of wildfowl. On application to the Holme Bird Observatory Warden, access is available to the observation hides overlooking the wader pools.

18. HOLME DUNES NATURE RESERVE (NORFOLK)

The reserve consists of a variety of habitats, including sand dunes, saltmarsh and foreshore. They attract a variety of birds both in the breeding season and throughout autumn and winter, when wader numbers are good. An interesting selection of migrants are recorded each year.

The 200-hectare (500-acre) reserve, managed by the Norfolk Naturalists' Trust, is rich in plant life, with sea buckthorn and sea holly abundant in places. Other plants include sea bindweed, sea lavender and two species of orchid – the bee and pyramidal orchids.

Access to the reserve is either by walking along the Thornham sea wall or by car from Holme village. Car parking facilities are available, but visitors should first contact the Warden at The Firs, Holme-next-the-Sea, Norfolk (Tel: Holme 240).

19. SNETTISHAM (NORFOLK) RSPB RESERVE

Shingle beach, saltmarsh, sand and mudflats and disused shingle pits covering 1,280 hectares (3,200 acres) form this RSPB sanctuary on the east side of the Wash, 3 km (2 miles) west of Snettisham Village.

The beach is approached via a minor road sign-posted from the A149 King's Lynn to Hunstanton road, and the reserve lies to the south of the public beach car park.

Over 70,000 waders and 10,000 wildfowl have been recorded on the reserve at any one time. Up to 170 bird species are sighted annually. These include knot, grey plover, turnstone, sanderling, common tern, Brent geese, various ducks and grebes.

Plant life supported on the shingle beach includes yellow horned-poppy, several oraches and sea sand-wort, rocket, beet and kale.

Four hides have been sited at the most southern shingle pit and can be visited all year free of charge. Further information is available from the Warden, School House, Wolferton, King's Lynn PE31 6HD.

20. GIBRALTAR POINT (LINCOLNSHIRE)

Covering a total of 420 hectares (1,050 acres) in all, this reserve is by far the most extensive area of sand dunes and saltmarshes on the Lincolnshire coast where accretion is still in progress. Every stage in the development of dune vegetation can be seen, cul-minating in an area of dense sea buckthorn. Plants include sea holly, shrubby seablite and sea bind-weed.

Nesting birds include little tern and ringed plover on the shore, together with a variety of small pas-serines in the dune scrub. Winter brings in large numbers of passage migrants and northern visitors such as fieldfare, twite and snow bunting.

Access to this huge reserve is via the Drummond Road from Skegness, which leads to Gibraltar Point, where good parks exist including one close to the visitor centre which is open daily from May to October together with the occasional weekend at other times of the year.

21. SALTFLEETBY-THEDDLETHORPE   DUNES (LINCOLNSHIRE)

This part National Nature Reserve is managed by the Lincolnshire and South Humberside Trust by agreement with the Ministry of Defence and Nature Conservancy Council. The land is leased from Lin-colnshire County Council and extends to some 475 hectares (1,185 acres) in all, and contains several types of habitat, including mud flats, sand dunes and salt and freshwater marshes.

Migration time is excellent for observing birds, and winter too brings in large numbers of waders and the occasional short-eared owl and hen harrier. It is the only site in Lincolnshire where the natter-jack toad is found.

Plant life is varied due to the range of habitats available on the reserve. In the dunes, both pyra-midal and bee orchids grow; the freshwater marsh contains sea rush and pond sedge; whereas in the more open areas early and southern marsh orchids as well as the rare marsh pea can be found.

There are five main access points from the main A1031 coast road. These are at Sea View, Rimac, Coastguard Cottages, Brickyard and Crook Bank. There are parking places at each entrance.

22. DONNA NOOK – SALTFLEET   (LINCOLNSHIRE)

This is an extensive reserve covering some 960 hec-tares (2,400 acres) or so, composed of sand dunes, mud flats and intertidal areas. The reserve is man-aged by the Lincolnshire and South Humberside Trust for Nature Conservation by agreement with the Ministry of Defence.

Breeding birds include ringed plover, little tern and oystercatcher. In winter, Brent geese and shel-duck are regular visitors, as are many wading birds. Passage migrants recorded include many rarities. Both grey and common seals can be seen at certain times of the year.

Botanically, the reserve is very interesting, with yellow-wort, bee orchid and pyramidal orchid growing on the stable dunes.

Access to the reserve is off the main A1031 coast road between Grainthorpe and Saltfleet, with park-ing areas at Stonebridge, Howdens Pullover and Sea Lane. Part of the area is a bombing range and under no circumstances should visitors enter the bombing area when the red flags are flying.

23. SPURN PENINSULA (HUMBERSIDE)

The Spurn Peninsula reserve is composed of a 5.5-km (3½-mile) long sand and shingle spit, managed by the Yorkshire Naturalists' Trust. In all, there are some 300 hectares (750 acres), extending from Kilnsea to the tip. The Humber estuary side is salt-marsh and mudflats, on which numerous waders and wildfowl feed at low tide – numbers being very large throughout winter.

From a narrow beginning, no more than 50 metres wide, the peninsula widens towards the tip, where stable dunes exist. Sea holly grows alongside several more common plants, especially sea buck-thorn.

Spurn is an excellent place to observe migrant birds and the bird observatory there is well situated for recording the several rare species that are regu-larly seen in the area.

Take the A1033 Hull to Patrington road and continue along the B1445 to Easington, from where unclassified roads lead to the headland. There is a charge for motorists entering the reserve. The car park near the lighthouse is quite close to Spurn's tip.

Ythan Estuary.

24. SANDS OF FORVIE AND YTHAN ESTUARY (GRAMPIAN)

An area of some 1,000 hectares (2,500 acres) cover-ing the sands of Forvie and the Ythan Estuary forms a National Nature Reserve, 19 km (12 miles) north of Aberdeen. The reserve is approached from the A975 road and lies between the towns of New-burgh, Ellon and Collieston. The sands of Forvie form one of the largest sand dune systems in Britain and the reserve is a fine example of coastal heath-land. The estuary is composed mainly of sand and mudflats.

Over 200 bird species have been recorded with over 40 of these being regular breeding species. The reserve has a large colony of breeding eider (6,000), four species of tern – sandwich, little, arctic and common; shelduck and ringed plover. The north-ern cliffs support kittiwake, fulmar and razorbill amongst other seabirds. Wintering wildfowl include greylag and pinkfoot geese, mallard, teal, golden-eye, long-tailed duck and whooper swan; waders include golden plover, greenshank and dunlin.

More than 340 plant species have been recorded in a variety of habitats: sea rocket and saltwort on the sand; wild pansy, white flax and violets in the dunes; crowberry, lousewort and heath milkwort in the moorland areas in the northern part of Forvie Sands; sea campion, kidney vetch, lovage, butter-

wort and grass of Parnassus on the heath slopes next to the dunes.

Fourteen butterfly species have been recorded and these are mostly seen on the cliff slopes. They include small pearl-bordered fritillary. Of the 200 recorded species of moth, day-fliers include the bur-net and wood tiger moth.

The reserve is open to visitors throughout the year. It can be viewed easily from the road and there are several good parking spots with viewpoints. An observation hide overlooks the tern colony. Infor-mation boards concerning the reserve are located at the main entrances to the dunes. Access to the southern end of the sands is not permitted between April and August due to the colonies of breeding terns nesting there. Further information is available from the Warden, Little Collieston Croft, Collies-ton by Ellon (Tel: Collieston 330 or 352).

25. CULBIN SANDS (GRAMPIAN/ HIGHLAND) RSPB RESERVE

Culbin Sands lie on the foreshore of the Moray Firth. Sandflats, saltmarsh, shingle bars and spits and a large sand dune system which has largely been covered with forest provide the principal habitats.

This RSPB reserve is approached along the shore

Culbin Sands and dune system.

Uists of the Outer Hebrides, where machair takes up 8 per cent of the total land area. It provides a light lime rich soil for the farmers of the area to cultivate, in sharp contrast to the acid peaty moorland covering the remainder of the islands.

The abundance of flowers in June and July makes the machair a breathtaking place to visit. Over 60 species of plants grow in this unique habitat, usually carpeting large areas for several months. Most are common species like daisies, storksbill, wild pansy and corn marigold; there are few rarities to speak of. What is surprising is that the machair is very light on butterflies, with the common blue being the only species likely to be seen.

The bird life is very rich, however, with breeding populations of ringed plover, redshank and dunlin well established, as well as the more common waders, the oystercatcher and lapwing. In the machair lochs, the red-necked phalarope and black-throated diver can be seen. Although the corncrake is rare in other parts of Britain, it is relatively common in the machair areas, although it prefers to nest in hayfields and marshes outside the true machair.

by taking the minor road by the golf course about 1.5 km (1 mile) to the east of Nairn. There is no visiting charge.

Common and velvet scoters, little and common terns, bar-tailed godwit, knot, dunlin and capercaillie include some of the many bird species regularly sighted here. Roe deer and badger are plentiful in the forest.

More information is available from the RSPB Scottish Office, 17 Regent Terrace, Edinburgh EH7 5BN (Tel: 031 556 5624).

26. THE MACHAIR (WESTERN ISLES)

Machair is best described as part of a coastal dune system originating from shell sand. It is a habitat that makes up a large percentage of the Outer Hebrides, although it does occur elsewhere in the Western Isles and there are small fragments in other parts of Britain too.

By far the greatest developments are along the

27. SOUTH WALNEY (CUMBRIA)

The 92 hectares (230 acres) at the southernmost tip of Walney Island constitute a nature reserve managed by the Cumbria Trust for Nature Conservation. Basically, it is a dune-covered shingle spit, much disturbed by gravel working. Nevertheless, the reserve provides a wealth of varied habitats including mudflats, saltmarsh, dunes, freshwater and brackish pools, in addition to the shingle.

Along with seabirds, many waders and wildfowl frequent the area and it is an excellent place to observe migrant passerines. In summer, South Walney is noted for the sizeable colonies of herring and lesser black-backed gulls it holds.

On the shingle surrounding the reserve, oyster-catchers and ringed plovers nest, and near the lighthouse four tern species (arctic, common, little and ringed) nest in tightly packed colonies, with numbers varying appreciably from year to year. Around them grows a typical shingle flora with an impressive display of viper's bugloss from June to September.

A special permit is required to visit this reserve, but these are available on site from the warden

Shingle and flowers at Walney Island.

based at Coastguard Cottages, Walney. The reserve is approached by turning left at the traffic lights after the bridge-crossing to Walney Island. Follow the Promenade and Ocean road and take the fourth road on the left, known as Carr Lane. Continue for 8 km (5 miles). Parking is available at Coastguard Cottages. An illustrated booklet on South Walney is available from the warden.

28. AYRES VISITOR CENTRE (ISLE OF MAN)

The Ayres is a stretch of shingle, dune and heath along the northern coast of the Isle of Man, culminating at the Point of Ayre. It is a good example of a raised beach and can be regarded as the 'newest' part of the Isle of Man. About 5,000 years ago, the

level of the sea became lower and the beach was built outwards from the cliff-line. Material from the nearby Jurby cliffs built up a series of spits or shingle ridges which developed from Blue Point towards the Point of Ayre. Wind-blown sand has accumulated on top of the shingle over the years.

The plants and animals found in the area today show the natural succession that has occurred, from the long colonised section inland to the newly colonised beach nearer the high-water mark.

The Visitor Centre is administered by the Manx Nature Conservation Trust and has displays explaining in detail the geology and natural history of the area. A warden is present throughout the summer months and is available to help with any queries. A nature trail exists to introduce visitors to the ecology of the area. This can be reached by following the A10 north-westwards from Bride, then turning right along a minor road some 1.2 km ($\frac{3}{4}$ mile) or so north-west of Bride village, signposted Ballaghennie. The first stop on the trail is the Visitor Centre itself which is easily reached on foot, along a concrete path which starts near the Forestry Board picnic site. This is an excellent place to look inland

and take in the whole sweep of the Ayres from the lighthouse to Blue Point, some 4 km (2½ miles) away. The Trail is well marked with concrete posts and coloured guide posts.

The Ayres is noted as an important breeding place for three species of tern. These nest on the upper shore and visitors are asked to avoid disturbing them during the breeding season. Dogs must be kept on a lead at all times.

Plant life is varied with marram grass dominating the sand dunes. Other species found there include sea holly, sea spurge and sea bindweed. On the more stable structures, wild thyme and bird's foot trefoil are quite common and there is a good variety of mosses and lichens. Several uncommon orchids grow in the damper dune slacks.

Further information can be obtained from the Manx Trust for Nature Conservation, 15 Athol Street, Douglas, Isle of Man.

29. LYTHAM ST ANNE'S (LANCASHIRE)

The nature reserve at Lytham St Anne's lies to the south of the railway line and to the east of the Pontin's Holiday Camp. It is essentially a system of sand dunes with two slacks frequented by bathing seagulls. The reserve is open at all times and there is no visiting charge.

Bird species include stonechat, linnet, skylark and kestrel. Rabbits, voles, mice and shrews abound and butterfly species present include the grayling, small heath, small copper and green-veined white. The reserve is important for moths – the white satin, cinnabar and burnet can be seen during the day and there are five species of bumble bee. Hot sunny weather brings out numerous common lizards.

Some 200 different plants grow on the reserve including early forget-me-not, heartsease pansy, early marsh orchid, round-leaved wintergreen and evening primrose.

A warden is regularly on the reserve. Based at the warden's hut just off the main road, he carries out routine management, compiles all scientific records and is available to help the general public to understand and enjoy the nature reserve.

The reserve is managed by the Lancashire Naturalists' Trust and more information is available from their headquarters at Dale House, Dale Head, Slaidburn, Clitheroe BB7 4TS.

30. AINSDALE SAND DUNES (MERSEYSIDE)

More than 680 hectares (1,700 acres) of the sand dunes between Hightown in the south and Southport in the north are nature reserves. The Ainsdale National Nature Reserve lies to the west of the railway between Formby and Ainsdale. The reserve provides a good example of a sand dune system and other habitats include marshy slacks, pinewoods and beach.

Recorded bird species include blackcap, greenshank, bar-tailed godwit, redpoll and treecreeper. The reserve is important for supporting colonies of the rare natterjack toad and sand lizard. Plant life is varied, with sea buckthorn plentiful.

There are no car parks on the reserve but parking facilities exist at Freshfield Station and the beach at Ainsdale. Ten km (6 miles) of marked pathways run through the reserve, including the Fisherman's Path from the station car park to the beach. In early summer, a nature trail, booked in advance, operates for school parties.

One interesting project undertaken on the reserve in recent years has been the removal of large areas of sea buckthorn, which tended to smother other plant life. The resulting bare patches now support an interesting flora.

Additional information is available from the Nature Conservancy Council, North West Region, Blackwell, Bowness-on-Windermere, Cumbria.

31. CEMLYN (GWYNEDD)

Situated on the north-western coast of Anglesey, close to the Wylfa Nuclear Power Station, this 14-hectare (35-acre) site is a nature reserve managed by the North Wales Naturalists' Trust on lease from the National Trust.

The shingle ridge is a good example of a 'bar' formed from storm-beach shingle. It holds a variety of interesting flora including sea-kale, vernal squill and sheep's bit.

The brackish pool behind the ridge is alive with common, arctic and sandwich terns during the breeding season, choosing to nest on a few small islands in the main pool. Some 700 or so pairs have been known to breed on the reserve, but numbers vary considerably from year to year. In recent years, 200 – 300 pairs are more commonly seen, mainly

Shingle flora along the ridge at Cemlyn.

due to the reformation of a breeding colony on the 'Skerries', a group of small rocky islands just off the mainland.

Winter visitors include such wildfowl as wigeon, shoveler and goldeneye with occasional visits from whooper and Bewick's swans. Other birds nesting on the reserve include oystercatcher and ringed plover which prefer the shingle towards the western end of the ridge.

There are good parking facilities at both ends of the reserve. To protect breeding species, visitors are asked not to use the public right-of-way along the ridge during the nesting season, but to keep to the seaward side along the high-water mark. To reach Cemlyn, turn right just after the entrance to Wylfa Power Station off the A5025 Amlwch to Holyhead road at Tregele. The reserve is several kilometres along this road.

32. RHOSNEIGR (GWYNEDD)

Visitors to this small coastal village on the west coast of Anglesey will not fail to be impressed by its superb sandy beach, the rich flora of the adjoining dunes and the variety of seaweeds and invertebrates in the rock pools. There is something to interest everyone and the various habitats are sufficiently close together not to warrant extensive walking.

In the dunes, sea bindweed, wild pansy, sea holly and sea spurge grow alongside the marram grass, whilst in the damper slacks near the main A4080 road to the village, orchids put on a glorious show during July.

No visit to Rhosneigr would be complete without a thorough investigation of the numerous rock-pools on the northern side of the beach. If the tide is out, it is possible with some care to go right down to the oarweed area of the lower shore. Here superb

Burnet moths (*Zygaena* spp.) mating on early marsh-orchid (*Dactylorhiza incarnata*) in a dune slack at Rhosneigr.

examples of sea belt and tangles, attached to the rocks, sway with the current. The walls of the pools are lined with brilliant red beadlet anemones, various periwinkles and limpets. Further up the shore, yellow, black and green lichens cover the rock and, in the grassy shingle area, thrift and scurvy grass grow around nesting oystercatchers in late spring and summer.

33. NEWBOROUGH WARREN (GWYNEDD)

Newborough Warren is part of an extensive National Nature Reserve extending to some 640 hectares (1,600 acres) in all, situated on the south-west coast of Anglesey. The dunes and grassland of the warren itself take up about 400 hectares (1,000 acres).

Severe storms in the early fourteenth century blew vast amounts of sand from the offshore banks, which built up into the present dune system. In 1948 the Forestry Commission began planting mainly Corsican pine to create what is now known as Newborough Forest. This was deemed necessary at the time because of the problems wind-blown sand caused to roads and agriculture in the area.

Marram grass dominates the mobile dunes near the sea, but other plants which are fairly common include dune pansy, sea spurge and mouse-ear chickweed. Inland, where the dune system is quite stable, lady's smock, tormentil, meadow saxifrage, bird's foot trefoil and wild thyme are found.

In the hollows or dune slacks, more moist conditions prevail, enabling creeping willow, buck's horn plantain, butterwort and grass of Parnassus to survive.

The reserve once supported nesting colonies of terns and various gulls, but nowadays apart from the occasional herring gull, oystercatcher, skylark and meadow pipit, the importance of the site as a nesting area is somewhat diminished.

Newborough Warren can be reached from the east by turning left off the main A5 trunk road into Llanfair P. G. and taking another left turn along the A4080 on entering the village. Continue along this road until the village of Newborough is reached. Turn left at the crossroads and continue along the road to the large Forestry Commission car park. Good toilet facilities are available.

Other routes are available for walkers; these are listed in an informative brochure available from the Regional Officer, Nature Conservancy Council, Penrhos Road, Bangor, Gwynedd. A written permit (available from the above address) is required to visit places away from the marked routes.

34. MORFA BYCHAN (GWYNEDD)

The mobile dune system nearest the shore is dominated by marram grass. A greater variety of wildlife exists on the fixed dune grassland behind, which gradually gives way to a freshwater marsh. Coastal plants include sea holly and the superb sharp rush, a very localised species found mainly in the west.

North Wales Naturalists' Trust members are able to park in the golf club car park and take one of the two routes across the links to the north-east corner of the reserve. Otherwise access to this 11-hectare (27-acre) site is gained by taking the road from Morfa Bychan to the beach car park, turning left and crossing the stream to the reserve entrance.

35. YNYSLAS DUNES (DYFED)

The dune system at Ynyslas, near the village of Borth in Dyfed, forms a major part of the Dyfi National Nature Reserve. In the embryo dunes above the storm-beach grow such plant species as sea rocket and prickly saltwort, whereas the fore-dunes are dominated by the stabilising marram grass with the occasional tough growth of sea spurge.

Running parallel to the fore-dunes are the more stable 'yellow dunes' and although marram is still the dominant plant other species have developed, such as red fescue and sand sedge. Lichens and fungi are also found in some places.

Rabbits are numerous and cause erosion problems as well as restricting the growth of some plants, especially orchids, whose flowering heads are nibbled away.

Butterflies seen on the reserve include the large skipper, meadow brown, wall brown, grayling and in some years the clouded yellow. Cinnabar and burnet moths are common together with the various banded snails.

The orchids of the dune-slacks are a major attraction and have flourished in the last twenty years or

Dunes at Ynyslas National Nature Reserve.

so. Several species of marsh orchid can be found in summer together with the marsh helleborine, bee orchid, pyramidal orchid and both common and heath spotted orchids.

Birds breeding in the dunes include skylark, meadow pipit, stonechat, linnet and wheatear. Some shelduck nest in disused rabbit burrows.

The well-stocked information centre on the reserve is open from April to September. A charge is made for parking on the beach which is easily reached from the village of Borth by turning off the B4353.

36. FRESHWATER WEST (DYFED)

Situated south-west of the town of Pembroke, Freshwater West consists of about a mile of scenic seashore and sand-dune system, owned by the National Trust. The area is relatively quiet even in summer, although it is a popular wind-surfing bay. Bathing is dangerous, however, owing to strong offshore currents and quicksands. Warning notices clearly emphasise this fact.

The beauty of the sandy shore is exemplified by the superb red sandstone rock structure to the north of the beach. Indeed the walk north-westwards along the Pembrokeshire Coast Path to West Angle Bay is refreshingly beautiful and well worth the effort.

Rock samphire and rock lavender flower on the

View over Freshwater West from the coast path.

rocks in summer and seaweeds of all kinds abound on the sandstone cliffs. The dunes hold an interesting flora, a large rabbit population and a variety of insects. Skylarks and meadow pipits nest in the marram grass.

The best approach is through the village of Castlemartin, which is reached by turning left off the B4320 Pembroke to Angle road. Continue along the B4319 to Freshwater West. Numerous parking places are provided by the National Trust, including clean toilet facilities.

Reserve, part of which is owned by the National Trust. Being exposed to the prevailing south-westerlies, some interesting hollows and patterns exist in the dunes.

The area is renowned for its variety of sand-loving plants and is little disturbed by people owing to its situation on the North Gower coastline. The roadway to the reserve is not suitable for heavy traffic and only a few smallish caravan sites exist in the locality.

Meadow pipits and skylarks frequent the dune grasses in summer. Rabbits are abundant, having little difficulty burrowing into the soft sand.

The route to the site is via the villages of Llanmadoc and Cheriton. With the south Gower coastline being completely full up throughout most of the summer, this part of the coast will offer visitors vast areas to explore and, of course, golden beaches.

37. WHITEFORD BURROWS (WEST GLAMORGAN)

This sand dune system extending to some 800 hectares (2,000 acres) is situated on the north-west tip of the Gower Peninsula. It is a National Nature

38. OXWICH NATIONAL NATURE RESERVE (WEST GLAMORGAN)

This reserve covering some 216 hectares (540 acres) in all lies 14.5 km (9 miles) west of Swansea on the

Sand dunes at Whiteford Burrows with the Worm's Head peninsula in the background.

Gower coast. It is approached by following the A4118 from Swansea, then after the village of Penmaen, taking the minor road signposted Oxwich. The beach is backed by sand dunes, beyond which freshwater and salt marshes have formed. The dunes have developed over the last 2,500 years and the stages in dune formation can be clearly seen. Marram grass has been planted and screens built to prevent sand erosion in some areas. Signs help to divert people away from these sensitive areas.

The reserve provides a variety of plant and animal life. The marsh behind the dunes supports waterlilies and bulrushes and dragonflies are numerous. Reed warblers also breed there. In winter, waders and ducks are plentiful.

Oxwich National Nature Reserve.

Woodland covers the dunes furthest from the shore and signposted footpaths exist throughout the area, including a nature trail through the dunes. Most of the reserve is open to the public and an information centre and car park is provided. Further information is available from the warden based at the Oxwich Reserve Centre, Oxwich, Swansea SA3 1LS.

39. KENFIG DUNES (MID-GLAMORGAN)

The area is perhaps better known for the 28-hectare (70-acre) pool situated near the village of Kenfig. The pool and dunes make up the Kenfig Burrows Nature Reserve, which is administered by Mid-Glamorgan County Council.

A visitor centre provides details of the plants and animals found on the reserve, together with information on recent sightings. Good parking facilities are available at the centre.

The pool is most interesting in autumn and winter, when passage waders and wildfowl numbers can be quite high. The dunes are noted for their range of flora, including an impressive variety of orchids such as twayblade, marsh orchid and the rarer bee orchid. A good variety of insects inhabit the area in summer, especially butterflies.

Visitors should leave the M4 at junction 37 and aim for the village of Kenfig where access to the reserve centre is well signposted.

3
ESTUARIES

Introduction

In simple terms, an estuary is a stretch of water at the mouth of a river where freshwater runs into the sea. However, since there can be no clear boundary between freshwater and salt water, there is a zone of brackish water, sometimes extending for many kilometres inland, the salinity of which varies with the changing tides.

One of the ways in which estuaries differ is the method by which the fresh and salt water running into them mix. In most cases, salt water, being denser, moves upstream along the bottom of the channels, allowing the less dense freshwater to remain on top. Even then there is some mixing, of course, which will depend on other factors as well, like the state of the sea and river at a particular time. On the Severn and a few other estuaries, however, the different waters mix thoroughly as the tidal wave moves up the estuary. Such a phenomenon has given rise to the famous 'Severn bore', a constant challenge for many surfers in their quest for the glory of travelling the furthest upstream.

In general, estuaries are usually well sheltered, especially those formed after the last ice age, some 10,000 years ago. Then, deep river valleys were drowned as the level of the sea rose. These rias are common along the coast of south-west England. In Scotland, fjord-like sea lochs were formed as a result of the drowning of valleys deepened by glaciers. In flat East Anglia and parts of north-west England, the rise in sea level had the effect of creating wide estuaries that extend considerably inland.

One noticeable feature of most estuaries is the way tidal mudbanks develop along their reaches. The sediment is partly brought downstream by the river, but much of it is carried upstream by the tide, having originated from erosion somewhere else along the coast. Much of the sediment is constantly moved up and down the estuary by strong tidal currents. This means that estuarine water is very turbid, limiting the species of plants and animals to those able to tolerate such conditions. During periods of heavy rain, whole mudbanks may be washed away, usually to be deposited downstream where the current is weaker. On top of all of this, there is the added problem of temperature extremes to cope with. Mudflats regularly freeze in winter and their dark colour results in rapid heating during periods of hot weather. Such a hostile environment is hardly conducive to wildlife, but the hardy species that can survive are normally present in dense populations.

The upper reaches of mudflats may be stabilised by the development of saltmarshes. These provide an important habitat for a variety of plants and animals. Some saltmarshes are extensive, like those of the Wash, which spread over some 4,000 hectares (10,000 acres) and form the largest almost continuous saltmarsh in Britain.

Both the mudflats and saltmarshes of our estuaries are of immense conservation importance and of international standing. In particular, they provide sites for several million migrant waders and wildfowl to roost and feed during the autumn to spring period.

Like other coastal habitats, estuarine areas are prone to severe disturbance and whole-scale damage by man. It would be reasonable to assume that they suffer more than any other, with one of the chief contributors to their destruction being the unending need for land reclamation. This is the process by which marshes and flats are drained to provide development land for agriculture or industry. Unfortunately, in most schemes, it is the outer mature marshland that is lost, by far the most valuable estuarine habitat for wildlife. It has been estimated that over twice as much saltmarsh existed around the coast of Britain only 500 years ago, and it seems that the east coast of England has suffered the most.

Energy requirements in the future have prompted much discussion into the possibility of barraging Britain's major estuaries. Should plans go ahead, massive reclamation would be necessary and there would be far-reaching problems caused by changes in the tidal flow, salinity and sediment deposition.

Pollution is not a new problem and will continue to plague our shores for generations more. In terms of sheer volume, domestic sewage creates the biggest threat. Its effect leads to the growth of green algae over the mud under which invertebrates would normally thrive, ultimately smothering them.

Industrial pollutants exist in many forms, ranging from toxic metals to organic chemicals. Several of our larger estuaries are used extensively by shipping or have oil refineries located on their shores. The introduction of oil into the estuarine waters, either accidentally or indeed deliberately, is a natural consequence, unfortunately. Fertilisers and pesticides washed from agricultural land upstream also pose a significant threat.

Direct disturbance exists in the form of recreational pursuits. Fishermen dig for bait on a large scale in some areas. Although recent inclusions in the Wildlife and Countryside Act limit the practice, much damage has already been done. They also disturb roosting and feeding birds, although the blame cannot entirely be put on them, as other individuals like wildfowlers and water-sport enthusiasts are also guilty at times.

Some form of increased control is required in the future to prevent further erosion of the estuarine environment. By far the most damaging threat is the siting of huge industrial complexes on the shores of estuaries; these often require additional land which is reclaimed from the neighbouring saltmarsh. More rigorous planning is essential, preferably with the co-operation of the various conservation bodies. Some estuaries have already been designated Sites of Special Scientific Interest, either in part or in full, primarily the ones requiring the most protection. Nevertheless, in areas like Teesmouth, protection now would be too late, and plans for

Looking across Torrisdale Bay from Torrisdale towards Bettyhill – an important wintering ground for wildfowl.

development along other estuaries are at such an advanced stage that the whole-scale ruination of vital coastal habitat is inevitable.

There is also a need for some radical thinking into new ways of controlling the amount of pollution entering the system, with possibly the creation of a government body just to watch over established wildlife sanctuaries which, at present, have no protection. Officers would be mobile on both land and sea. The need to protect existing unspoilt estuaries and to improve conditions on the others is of paramount importance if the natural life of the coast is to be enjoyed by future generations.

Flora and Fauna

The range of species of aquatic plants and animals found in estuaries is limited and contains only those that are able to tolerate the constant and rapid changes in salinity, turbidity and, to some extent, temperature that occur throughout the twice-daily tidal cycle. The species that do exist in this hostile environment, however, occur in huge populations, making an estuary one of the most productive natural ecosystems on earth.

So little light passes through the murky waters that few seaweeds survive. One exception is the green *Enteromorpha*, which often covers large areas of mudflat and is particularly tolerant of fresh water. It forms a major part of the diet of several ducks and geese. Although the main concentrations of wracks are found on rocky shores, the horned wrack does frequent those estuaries where there are plenty of attachment points for its branched holdfast, like rocks, stones, timbers and other structures.

Unlike the wracks, eelgrass thrives without the need for additional support. Its deep roots and extensive underground stem give it the necessary stability to put up with strong currents whilst also helping to stabilise vulnerable mudbanks.

The unlimited supply of nutrients from river and sea is essential for the development of a saltmarsh plant community. Amongst the first plants to colonise a mudflat will be marsh samphire or glasswort, and probably common cord-grass. Where common saltmarsh grass develops, hummocks are created because of its ability to trap sediment so easily. On the more established sites, more colourful species like sea aster, sea thrift and sea lavender grow alongside the more typical fleshy plants of estuaries like seablite, sea purslane and orache. Apart from producing attractive flowers during August and September, sea aster also plays host to the star-wort

moth, whose handsome green and black caterpillars feed on the plant.

In the nutrient-rich estuarine waters, plant plankton also thrive, providing food for filter-feeding invertebrates living in the mud surface, like cockles and mussels, which in some areas occur in such large concentrations that they provide a living for the shellfish-gatherers. There are also the shrimps and worms, which are just as well-adapted for tolerating the hostile environment, being able to burrow deeply into the mud or sand when conditions on the surface deteriorate.

By far the most numerous estuary mollusc is the laver spire shell, often referred to by its generic name 'hydrobia'. Population densities can exceed 30,000 per square metre and they tend to be concentrated near the surface of the mud. There they will be preyed upon by the shorter-billed waders like the plovers, which will also eat other surface species. The shelduck's surface-sieving feeding technique is also well adapted for taking large numbers of these tiny gastropod molluscs. Less numerous than hydrobia, but still present in densities of up to 6,000 per square metre, the bivalve Baltic tellin prefers clean sand and is also preyed upon by birds.

Waders with a moderate bill-length, such as dunlin, knot and redshank, tend to probe to a depth of up to 4–5 cm ($1\frac{1}{2}$ – 2 in), and in doing so are likely to encounter many worms and bivalves as well as the amphipod crustacean *Corophium volutator* and the brown shrimp (*Crangon crangon*).

The deeply burrowing lugworm and ragworm are mainly preyed upon by the larger-billed birds such as the godwits, curlew and pintail. The oystercatcher must be regarded as the main predator of molluscs like cockles, and because of the commercial value of these bivalves some conflict of interest

exists in parts of Britain where a cockle industry exists.

Wormcasts in the mud at Lindisfarne Nature Reserve.

Many mud-dwelling invertebrates remain hidden for most of the time, emerging only to feed during high water; but one estuarine creature that rarely conceals itself, except under weeds if present, is the common shore crab. It has the ability to survive in the upper reaches of estuaries where the salinity is extremely low, and lives on the abundant decaying material of the saltmarsh. The shore crab is itself preyed upon by birds like the curlew and turnstone.

It has been estimated that not far off two million waders and over a million wildfowl use the estuaries of Britain during the year. All rely on the highly-productive estuarine environment to provide them with nourishment, especially during the winter months when huge flocks of waders including dunlin, oystercatcher, knot, plovers and curlew feed on our shores, together with sea-duck, swans and geese. Other species also visit estuaries to fish, notably cormorants, grebes, herons and, mainly in the north, divers. Predatory animals include the otter, especially in Scotland, and both the common and grey seals, although all of these also frequent other coastal habitats.

Some birds nest in the upper reaches of estuaries, on or near the saltmarsh. Colonies of black-headed gulls breed in some locations, usually in large numbers. Shelduck nest in the denser vegetation on the edge of a marsh and frequently use disused rabbit burrows. Smaller birds like meadow pipits will also breed in the more mature areas. In the main, however, the unstable tidal nature of estuaries makes them generally unsuitable for nesting, especially when high spring tides coincide with the breeding season.

Several species of cord-grass are present on British estuaries, but one hybrid between *Spartina maritima* and *Spartina alterniflora* shows such vigorous growth that it must be considered as a damaging species. It has been deliberately planted in some areas to encourage reclamation, but it has the effect of severely limiting the growth of other saltmarsh species. Because it creates a dense vegetation, estuarine birds are less inclined to use spartina marshes, preferring more open situations.

Black-tailed Godwit (*Limosa limosa*)
Length 40.5 cm (16 in)

Of the two godwit species occurring in northern Europe this is slightly the larger. In spring and summer the back is mottled chestnut and dark brown; head and long neck are reddish-brown; belly and flanks are pale fawny-white barred with dark reddish-brown; rump white; tail black. Bill long and straight, mainly orange but shading dusky black towards tip. Legs long and slate-grey. During the breeding season the extent and richness of reddish-brown is less than in the bar-tailed godwit, *Limosa lapponica* (which incidentally does not breed in Britain). The two are readily distinguished in flight; the black-tailed having a prominent broad white wing bar, and legs that project well beyond tail. Both these features are lacking in the bar-tailed.

Since about 1952 after a long absence as a breed-

Black-tailed godwit (*Limosa limosa*).

ing species in Britain, the population has slowly but steadily increased to around 50–60 pairs, the main area being the Ouse Washes. Also breeds locally but sparingly on the Solway, Shetland Isles and parts of Yorkshire.

A noisy bird at all times but especially so on the breeding grounds; the loud disyllabic 'recka' of the male is characteristic; as is the loud and oft-repeated 'tur-ee-tur' during his display flight.

The black-tailed godwit nests in loose colonies; its four pear-shaped, olive-green to olive-brown eggs are spotted and blotched with darker shades of brown and scrawled ashy-grey. They are laid late April to early May in a ground nest which is often in thick tussocky grass. Incubation continues for approximately 24 days, a duty shared by both sexes. The young are equally cared for by both parents. On the breeding territory the most frequently heard calls are a nasal 'quee-it' and a rapid tittering 'tiu-i-tiu'.

Food is acquired by probing deep into mud and soft soil when insect larvae and worms are skilfully located by the sensitive tip of its long bill.

Breeding distribution: Iceland (an estimated 53,000 pairs mainly in the south-west, more sparingly in a few northern areas); Faeroe Islands (sparingly); Russia west of the Ural Mountains between latitudes 50°N and 60°N; Holland (sometimes around 50,000 pairs); Denmark; Belgium; France (near mouth of River Loire, and the Dombes northeast of Lyon – sparingly); parts of Germany and northern Poland. *L.l. melanuroides* replaces the nominate in Eastern Asia.

The main wintering quarters are the Mediterranean coast south to tropical Africa, the Nile Valley, also from the marshes of Iraq south to the Persian Gulf. It is estimated that approximately

70,000 birds now winter in Europe; this includes about 5,000 birds in Britain and Ireland (these are mainly on the south coasts, the exception being the Dee Estuary with 600–700 birds). However, around the coasts of Britain the black-tailed is far outnumbered by wintering bar-tailed, with peaks calculated at 50,000 birds (6,000–7,000 of which occur in Morecambe Bay).

There are two races: *L.l. limosa* of Europe and W. Asia and *L.l. melanuroides* of N.E. Asia.

Good areas to watch in winter are: Ribble and Dee estuaries; Stour estuary; Chichester Harbour; Exe estuary; Bridgewater Bay; Burrey Inlet (Glamorgan); and Eden estuary (Fife).

Curlew (*Numenius arquata*)
Length 53–58 cm (21–23 in)
Breeding birds often frequent boggy moorland and hill pastures; sometimes marshy fields, heaths, and sand dunes. The curlew, Europe's largest wader, has buffish brown plumage streaked with dark browns; the bill is long, slender and downcurved. Female

Curlew (*Numenius arquata*) approaching its nest.

smaller than male, but she has a slightly longer bill. The call is a clear ringing 'cour-li, cour-li' or 'croo-ee, croo-ee'.

During winter months rocky and sandy shores are frequented, as are estuarine areas. Some good places to watch are Morecambe Bay; Ribble and Dee estuaries; the Wash; Lindisfarne (Holy Island); south bank of Solway; Foulness and Maplin sands; Medway marshes and estuary; Torridge-Taw estuary; Humber estuary; and Burry Inlet (Glamorgan).

There are two races: *N.a. arquata* breeds N. Europe and Russia; *N.a. orientalis*, C. Asia.

Whimbrel (*Numenius phaeopus*)
Length 41 cm (16 in)
Upperparts mainly sandy brown, streaked and mottled with buff. The dark crown has one central and two outer buffish stripes (supercilia). Underparts buffish-white heavily streaked with brown. Bill shorter, much darker, and more abruptly downcurved than the curlew (*N. arquata*).

It breeds sparingly in northern Scotland, Lewis, and Shetland; favouring boggy moorland.

The flight call consists of four notes 'kwip-pip-pip-pip' or 'tetti-tetti-tetti-tetti'.

In winter months frequents seashores and estuaries; Bridgewater Bay in Somerset, the Wash and Blackwater estuary in Essex are excellent areas to watch.

There are three races: *N.p. phaeopus* of N. Europe and N. Asia; *N.p. variegatus*, E. Siberia; and *N.p. hudsonicus*, N. Canada.

Bar-tailed Godwit (*Limosa lapponica*)
Length 38 cm (15 in)
An Arctic breeding wader very resplendent in its full summer plumage which is mainly rich chestnut, mottled and streaked with brown. A winter visitor to our shores; now the rich chestnut has been replaced with mottled greys, and the underparts are whitish. The long bill is slightly upturned. Body size a little smaller than the straight-billed black-tailed godwit, *L. limosa* (length 40.5 cm, 16 in).

Very silent outside the breeding season. When feeding often wades belly-deep close to the water's edge, immersing the entire head as it probes the mud with that long bill. Also searches actively on beds of decaying seaweeds for insects.

The two races are *L.l. lapponica* of N. Europe and N. Asia and *L.l. baueri* of N.E. Asia and N.W. Canada.

Some of the best areas to see wintering birds are: Morecambe Bay, where they can number between 6,000 and 7,000 birds; the Ribble and Dee estuaries; Lindisfarne (Holy Island); the Wash; Firth of Forth; Inner Moray Firth; Eden estuary (Fife); and north Solway.

Over one-third of the bar-tailed godwit's breeding population winters in Britain.

Dunlin (*Calidris alpina*)
Length 18 cm (7 in)

A stocky and short-legged bird with a somewhat hunched posture. In the northern hemisphere it is the most numerous of all small wading birds on passage, or during the winter months, with an estimated population in excess of half a million in the estuaries of Britain and Ireland. A highly gregarious species at this time, often in company with knot (*Calidris canutus*) and other small waders. In winter its predominantly grey plumage and white underparts may cause confusion with the sanderling (*C. alba*) or the curlew sandpiper (*C. ferruginea*) (which is of similar body size but taller). The sanderling is larger and whiter, its distinctive white wing bar contrasting against the black of the primary coverts. The curlew sandpiper's conspicuous white rump and blackish tail make it readily distinguishable in flight. The winter migration extends it to almost every country down to 20°N.

In breeding plumage the Dunlin's reddish brown crown and upperparts are streaked with black; the throat and neck are greyish streaked with black; the white upper breast is also streaked with black; the lower breast has a large black patch (a diagnostic feature). Flanks are whitish. Iris black. The black bill is fairly long with a slightly decurved tip. Sexes are alike.

During the breeding season the dunlin shows a preference for moorland and tundra regions, but also frequents coastal marshes within its breeding range. The nest is well concealed under cover of thick low vegetation, a depression sparsely lined with grass. Four greenish brown eggs are laid; these are spotted and blotched with dark browns, especially so at the broad end. Often these markings appear to be slightly spiralled around the shell.

Laying usually occurs during April or May and the 21–22 days' incubation is shared by both birds. The dunlin's delightful purring trill once heard is seldom forgotten.

Five subspecies are recognised. The nominate *C.a. alpina* breeds from N. Europe to N.W. Asia; *C.a. arctica* breeds in E. Greenland; *C. a. centralis*, N. Siberia to Mongolia; *C. a. hudsonia*, the western Hudson Bay area; *C.a. pacifica*, N.E. Asia, Alaska, and N.W. Canada; *C. a. schinzii*, British Isles and Holland.

In winter months good numbers are in evidence at Morecambe Bay, Ribble and Dee estuaries, The Wash, and Chichester Harbour.

Knot (*Calidris canutus*)
Length 25.5 cm (10 in)

This small but plumpish wader is very resplendent in its breeding plumage, with black and chestnut upperparts, and uniform rich chestnut orange below. Alas, by the time they arrive on our shores from their summer homes in the high Arctic, most have moulted and are now ashy-brown above, with whitish underparts. The knot is a highly gregarious species, with probably the majority of the world population wintering in Britain and Ireland; estimated peak numbers are in the order of 300,000 birds. Visit one of their favoured estuaries (e.g. Hilbre in the Welsh Dee or Estuary and Morecambe Bay) and see how they pack so closely together, from a distance looking like a grey shimmering mass. In flight they retain this closeness, and when banking from side to side in unison the reflected light from their plumage suddenly changes from grey to silver.

Often they fly in company with dunlin (*Calidris alpina*), which of course they outnumber. Even though the latter is only 18 cm (7 in) in length it is not the easiest of tasks to separate them at distance. Knot are paler and larger, with blackish primaries and a none-too-distinct white wing bar. Look also for the whitish rump and tail; in both the dunlin and the sanderling (*Calidris alba*) these areas are uniformly darker.

There are three races: *C.c. canutus*, Spitzbergen, Taimyr Peninsula; *C.c. rogersi*, Siberian islands, E. Asia; *C.c. rufus*, Greenland, N. Canada.

Some other good wintering areas are Ribble estuary, Firth of Forth, Humber estuary, and south bank of the Solway.

Lapwing (*Vanellus vanellus*)
Length 30 cm (12 in)

By far the most common British wader with a breeding population in this country of about 200,000 pairs. Nests mainly in lowland areas, and to a much lesser extent in the uplands. Is it really necessary to describe the black and white plumage of such a familiar bird? Other features include a prominent long black curved crest, and that distinctive wheezy call 'pee-weet, pee-weet' as it engages in aerobatics, twisting, plunging and rolling above its territory; the wings producing a loudly pulsating 'lapping' sound.

Laying commences in late March continuing into April; three or four eggs are deposited in a shallow scrape either on bare ground or in a grassy situation. They are dark buff or olive, boldly blotched and spotted with blackish-brown. The female undertakes the greater share of incubation which continues for 24 to 26 days.

Lapwings begin to flock in June and by July the breeding territories will be deserted; it is now that a general southerly movement gets underway. In winter months large flocks will gather to feed on coastal marshes and mudflats especially during severe weather conditions.

The lapwing's breeding distribution extends from western Europe eastwards to China and Japan; there are no sub-species.

Redshank (*Tringa totanus*).

Enormous winter flocks can often be seen on the south bank of the Solway; other good places are the north Solway; Torridge-Taw estuary, Devon; Camel estuary, Devon; Morecambe Bay; also the Ribble and Dee estuaries. These are but a few suggestions.

Greenshank (*Tringa nebularia*)
Length 30 cm (12 in)

The total breeding population of the Scottish Highlands and islands is put at about 500 pairs; this is the western limit of the greenshank's range. A little larger than the common redshank (*T. totanus*), but without a white wing bar; and with whiter underparts.

In the main a grey bird with a white rump; it has a dark and slightly upturned bill, and long greyish green legs. Its singing call 'tew-tew-tew' is a little less shrill than the common redshank's.

With only one race described it is interesting to note that some birds that breed in the mire forests of Swedish Lapland have decidedly brownish plumage and greenish-yellow legs.

Some good places for watching in winter include the Colne and Blackwater estuaries, Essex; the Wash; south bank of the Solway; also the Medway marshes and estuary, Kent.

Redshank (*Tringa totanus*)
Length 28 cm (11 in) **Fig. 10(1)**

In the summer months, when breeding, the redshank moves inland to favoured areas of moorland close to reservoirs, and inland lakes, also water meadows. Coastal breeding is mainly restricted to saltings, shingle beaches with a little vegetation, and sand dunes. The British population is probably in the region of 43,000 pairs, give or take a few thousand either way. It is one of our best-known medium-sized waders and widely distributed; much more common in the north than the south, but far from common in parts of the south-west and Ireland.

When standing on those conspicuous orangey-red legs it is just a grey-brown bird with a long reddish bill. More easily identified in flight; now the almost completely white belly, the white rump, and broad white wing bar become apparent. Legs do not project beyond the tail. On alighting the character-

Fig. 10   1. Redshank (*Tringa totanus*). 2. Spotted redshank (*Tringa erythropus*). Birds in winter plumage.

istic habit of extending its wings upwards reveals their light under-surface.

The nest is usually well hidden under tussocky grass and lined with dry grasses. During late April or early May, four pear-shaped eggs are laid; these are buffish, copiously spotted and blotched with reddish brown. Both sexes share the incubation for 21 to 25 days. One of its better-known alarm calls is a strident 'tew-hew-hew', a call often uttered when perching is 'chip-chip-chip'. When in flight a repeated call of 'tut, tut, tut' may be heard.

Winter months are mostly spent in the vicinity of estuaries and mudflats around the coast.

Four races are identified: *T.t. robusta* from Iceland to W. Europe; *T.t. britannica*, British Isles to W. Europe; *T.t. totanus*, N. Europe and W. Siberia; *T.t. eurhinus*, C. and E. Asia.

The Wash, the Stour and Blackwater estuaries, Essex, Foulness and Maplin sands, Essex, Ribble estuary, south and north banks of the Solway, and Firth of Forth are just a few good areas to watch for redshank in winter.

Spotted Redshank (*Tringa erythropus*)
Length 30.5 cm (12 in) **Fig. 10(2)**
Breeds in northern regions of Europe and Russia. In summer its sooty-black plumage, with white spots on mantle and wings, makes it readily distinguishable from all other waders. When seen on British shores it is usually in winter plumage and looks very much like the common redshank (*T. totanus*).

If in company with each other the spotted redshank is obviously larger, and has longer legs. In flight two other features which help distinguish it from the common redshank are lack of white wing bar, and legs that project beyond tail.

Good places to look for the spotted redshank in winter include: Blackwater estuary in Essex; Dengie, Essex; also Medway marshes and estuary, Kent.

Pintail (*Anas acuta*)
Length drake 65 cm (25½ in), duck 55 cm (21½ in)
Scotland, northern and eastern England are the only breeding areas within the British Isles; and with this population estimated at less than fifty breeding pairs the pintail is indeed thin on the ground.

The drake's distinctive appearance makes iden-tification an easy matter, with chocolate-coloured head and neck, a narrow white band extending down sides of neck and joining the white breast. Back and flanks are very finely vermiculated with dark grey and white; belly white. The bronzy-green speculum has a tawny border in front and is white-edged behind. Tail long and pointed; legs and feet are grey. The female resembles a mallard duck, although more graceful with a longer neck, and a pointed tail; but the speculum is obscure.

When breeding, it frequents marshes, shallow freshwater lakes, and tundra pools, preferably if the surrounding areas are flat and dry.

Not very vocal, the drake has a low melodious double-noted whistle, and the duck a low quack.

The nest is loosely constructed of dry vegetation and not necessarily well concealed; it is lined with down and feathers. Clutches are found from May onwards, with replacements on to August. A normal clutch would be seven to nine greenish-buff or cream-coloured eggs. The female alone incubates and hatching follows after 23 days. During the six-week fledging period, although she tends to her ducklings, they must feed themselves.

In winter months they are more regularly found in coastal areas, less so inland. Good areas to watch include the Ribble and Dee estuaries; Stour estuary, Essex; Medway marshes and estuary, Kent; and north bank of the Solway.

The nominate *A.a. acuta* breeds in N. Europe, Asia, N. America, and W. Greenland. There are two isolated sub-species in the southern hemisphere: *A.a. eatoni* on Kerguela Island, and *A.a. drygalskii* on Crozet Island.

Shelduck (*Tadorna tadorna*)
Length 60–65 cm (23½–25½ in)
Somewhere in the region of 12,000 pairs breed within the British Isles, these almost exclusively in coastal areas. This breeding population possibly represents about half the total for all north-west Europe. Another interesting fact is that of the estimated 125,000 birds that winter within north-west Europe, 50,000 are to be found around the coasts of the British Isles, where the largest concentration of up to 14,000 birds has been recorded in the Wash.

The shelduck's large size gives it a goose-like appearance. When in breeding plumage the adult male has head and upperparts of dark glossy green;

this contrasts markedly with the white of lower neck, upper breast, and underparts. There is a broad chestnut band encircling the mantle and breast, and similarly coloured undertail coverts; scapulars and primaries are black; wing coverts white. Bill is upturned and red, with a basal knob on upper mandible. Legs and feet are pale pink. Female is noticeably smaller, with slightly duller plumage, and has no knob at base of upper mandible.

During March and April territory boundaries are established on the feeding areas close to the nest site. At this time listen for the characteristic chattering call 'ag-ag-ag-ag-ag'. The male also has a low whistling call, and the female a harsh barking 'quack'.

A typical nest-site would be down a disused rabbit burrow in a sand-dune or bank, with the nest of pale grey down constructed at a distance of 2 metres from the entrance. Other, less typical sites include holes under trees, also cavities in trees and walls. Eggs are laid from early May onwards; a normal clutch would be from 7 to 12 creamy-white eggs; nests found with larger numbers are often the product of two females. Incubation is by the female alone for a period of 28–30 days; she is called off to feed perhaps three or four times each day. On hatching, the parents lead the ducklings away to the water; this often involves walking some considerable distance, on occasions several kilometres. In areas where the breeding density is high many broods will join together, rather like a nursery group, under the watchful eye of just a few adults; the reason for this behaviour is not fully understood.

Shelduck feed principally on the small saltwater snails *Hydrobia ulvae* which live on the surface of estuarine mud, but at low tide burrow themselves just beneath the surface. They are eagerly sifted out by the 'dabbling' shelducks; so plentiful are these molluscs that concentrations of up to 60,000 per square metre have been recorded.

After breeding most western European shelduck migrate to the Heligoland Bight in West Germany to complete their moult. During August and September their numbers may build up to a staggering 100,000 individuals.

Apart from the Wash, which has already been mentioned, other good areas for watching during winter include: the Dee estuary; Chichester harbour, Sussex; Bridgewater Bay, Somerset; Poole harbour, Dorset; Langstone harbour, Hampshire;

Thames-side Marsh, and the Swale, Kent; Stour estuary, Essex; and Teesmouth, Cleveland.

The Shelduck's breeding distribution extends from W. Europe to E. Asia, and N. China. There are no sub-species.

Teal (*Anas crecca*)
Length 35 cm (13½ in)

This is the smallest of the ducks resident in Europe, and often its small size alone is sufficient means for identification. It is widely distributed throughout the northern hemisphere and one of the most numerous ducks. The breeding population in the British Isles is estimated at 4,750 pairs, give or take a thousand. Total for north-west Europe is about 150,000 birds, and the Mediterranean population is estimated at 750,000.

The drake has a rich chestnut head and neck, with a broad metallic green band running backwards from in front of the eye down towards back of neck; this band is narrowly bordered with pale buff. Mantle and flanks are evenly vermiculated pale grey and black. White patches on the bird's long scapulars produce a narrow band of white along the body above the wing. The black and green speculum has a white border in front and behind. The upper breast is palish cream and duskily spotted. Remainder of underparts are white, slightly barred on abdomen and vent. Central undertail coverts are black, and on either side of rump there is a creamy-yellow patch. Bill dark; legs and feet greenish-grey. The female has upperparts mottled and streaked with dark brown and a little buff; crown dark; underparts white. In eclipse plumage the drake resembles the female but has darker underparts. Call of drake is a high-pitched whistling 'krit', and the female a high-pitched 'quack'.

It is an inland nester associated with grassy or heather moorlands with areas of peat bog, low scrub, small ponds, and lakes. Eggs are laid from April to May; these usually number eight to ten, they are creamy-white with a greenish tinge and incubated by the female for about 21 days. The ducklings fledge at about four or five weeks old.

Shelduck (*Tadorna tadorna*).

During winter months the teal frequents lakes, ponds, flooded meadows, and marshes; however, at this time it is mainly coastal, visiting estuaries and mudflats.

Some good places to watch in winter are the Ribble estuary, Lancashire; Medway marshes and estuary, Kent; Newton Marsh, Isle of Wight; Dyfi estuary, Dyfed; Tay estuary, Tayside; and Cromarty Firth, Highland.

The nominate *A.c. crecca* breeds in Europe and Asia; and many Asian birds winter south in tropical Africa, Arabia, India and south-west Asia. *A.c. nimia* the Aleutian Is; *A.c. carolinensis* N. Canada.

Wigeon (*Anas penelope*)
Length 46–56 cm (18–22 in)

The total wintering population of north-west Europe and the Mediterranean is put at a staggering 1,000,000 birds or thereabouts; east of this the totals are not known with any accuracy. In Britain we have a breeding population of perhaps less than 500 pairs, most of which are found in the Scottish Highlands, and to a much lesser extent in southern Scotland and the Yorkshire Pennines; elsewhere the Wigeon must be regarded as but an occasional breeder.

The Wash, however, can boast the largest concentration of wintering birds in Europe, with peak counts of 42,000 recorded in January. Thanks for this important inland site are due to the Wildfowl Trust and the Royal Society for the Protection of Birds, both organisations having developed conservation areas thus minimising disturbance and improving both water and land management.

Several fairly conspicuous features make identification of the drake a relatively easy matter: a chestnut head with prominent creamy-yellow crown; pinkish-brown breast; a mainly grey body with a narrow white band at the edge of its wing; underparts mainly white contrasting sharply with the black of undertail coverts. Bill slatey-grey with black tip. Legs and feet greyish-brown. Unfortunately the female has no outstanding field characteristics, and at distance or in poor light may only be identifiable by the presence of a male. She is mainly brown above and white below; lacking the black undertail coverts, white wing patch, and the creamy-yellow crown. Look for the same general shape, particularly that high rounded head.

The nest is usually a simple affair in thick waterside vegetation, sometimes under the shelter of a low bush, and lined with down. Eggs are laid from early May onwards; they number seven to ten, are creamy-buff, and incubated for three-and-a-half weeks; the ducklings fledge at about six weeks old.

The drake has a melodious whistling 'wheeooooo'; the female produces a low purring note.

Wigeon feed exclusively on vegetable matter. They may be seen cropping the short grasses on watermeadows well inland, as at Welney Washes in East Anglia; or grazing on grass-covered saltings. They do, however, also feed extensively on mudflats where eel grass (*Zostera*) is the principal item along with the green algae *Enteromorpha*.

For coastal winter-watching, Lindisfarne (Holy Island); the Wash, Lincolnshire; Medway marshes and estuary, Kent; Cromarty Firth, Highland; Tamar estuary, Devon; Chesil Fleet, Dorset; the Swale, Kent; Stour estuary, Essex; and the Ribble estuary, Lancashire, are just a few good areas.

Its breeding distribution includes N. Europe and N. Asia. There are no sub-species throughout its range.

Mallard (*Anas platyrhynchos*)
Length 58 cm (23 in)

Surely the most familiar duck in Britain, and possibly throughout Europe; it is regarded as *the* wild duck in many countries. The mallard's estimated breeding population is put at 1,500,000 pairs, about half of which breed in the Soviet Union. The Netherlands, Finland and the British Isles each have upwards of 150,000 breeding pairs. During winter the resident breeding populations of Western and Mediterranean Europe are joined by birds that have bred further north in Arctic and sub-Arctic regions.

We doubt that the drake in breeding plumage needs any description, but even so, very briefly, he has a metallic green head and purplish-brown breast, these two areas separated by a very distinct narrow white neck band. The body is grey; the tail grey and white with tightly-curled black central feathers. The wings are brownish-grey and have a mirror of iridescent purple, bordered on each side by bars of black and white. Bill greenish-yellow, legs orange-yellow. The female has a uniform plumage of mottled buffish-brown; there is a dark brown stripe through the eye and a pale eyebrow. The wing mirror is less distinct than the male's. Bill olive-brown but mottled with orange towards edges. Males in eclipse plumage resemble a dark female but with a reddish-brown breast.

Courtship and pairing begins in October and, if the weather is mild, nest-building could be as early as February. Later nesters have the advantage of denser cover under which to conceal their nest, which is usually close to water. Sometimes nests are built in trees and very occasionally in holes. The 10 to 12 buffish-green eggs are laid on a bed of down. The female incubates for about 28 days, and her mate stays in the vicinity during the early stages; during this short period they may be seen feeding together. Soon he forsakes the sitting duck, who alone tends to the needs of the ducklings. From the outset they acquire their own food, and at about eight weeks old have fledged and become independent.

The drake's call is a quiet 'yeeb', and the female's a loud and very familiar 'quack'.

The nominate *A.p. platyrhynchos* has an extensive breeding range throughout the northern hemisphere and includes all but the most northerly regions of Siberia and North America. There are six sub-species: *A.p. conboschas*, Greenland; *A.p. wyrilliana*, Hawaiian Is; *A.p. laysanensis*, Laysan Is; *A.p. fulvigula*, S.E. USA; *A.p. diazi*, N. and C. Mexico; and *A.p. maculosa*, Mexico.

For winter-watching, the Ribble and Dee estuaries; the Wash, Lincolnshire; Bridgwater Bay, Somerset; and Tay estuary, Tayside, are just a few of many.

Red-breasted Merganser (*Mergus serrator*)
Length male 65 cm (25½ in), female 58 cm (23 in)
Fig. 11(2 and 2a)

Of the three 'sawbill' species which occur in the British Isles, only one, the red-breasted merganser, is a coastal breeder (although it too occasionally breeds inland). The goosander (*M. merganser*) chooses a habitat close to freshwater; as does the smew (*M. albellus*) which only visits Britain after breeding. Furthermore the red-breasted merganser is the only one to winter almost entirely in coastal waters (Fig. 11, 3 and 3a). The other two much prefer a freshwater existence, but during winter months when the larger lakes and other inland waters become frozen-over they too, compelled by

hunger, visit estuaries and sheltered coastal bays. The resident population in Britain appears to be about 2,000 pairs and increasing, the stronghold is western Scotland, with small numbers in northern Scotland and south to the Lake District, and it also breeds in Wales.

The drake is easily identified with that bottle-green head and long shaggy double crest, a wide white collar, and reddish-brown breast spotted and flecked with black. The outer scapulars and wing coverts are white edged with black; remainder of wing black with white patches on secondaries. At close quarters the white spots on its black shoulders become apparent. Undersides are creamy-white, the flanks finely barred with black. The long slender ruby red bill has a black central line from base to hooked tip, and is furnished with fine backward-sloping serrations; hence the name 'sawbill'. Legs and feet are some shade of red. The female has a reddish-brown head, and a double-tufted crest which is held out horizontally (the crest of the female goosander tends to droop). Her body is mainly barred and mottled medium grey and brown. Look also for merging of head and breast colours (female goosander has these sharply defined). Bill and legs as in drake.

It nests at ground level in dense scrub or heather, also under gorse bushes and brambles, and usually close to a loch, lake or river; down is added when egg-laying is underway. An average clutch would be eight to ten greyish or greenish-buff eggs; the time is late May to June. Incubation is by the female and takes four-and-a-half weeks; the drake only rejoins his mate when the young have hatched and together they tend to their ducklings.

When searching for food often swims with head submerged, or explores the underwater terrain by probing crevices with bill. Fish form the bulk of its diet and now the value of those serrated mandibles is seen to good effect.

The drake is usually silent other than during his display when a low mewing 'yeoww' will be heard; the female has a harsh croaking call 'krrrr'. In flight both sexes utter a call described as 'wark'. Although they often do fly at height it is perhaps more usual to see them travelling low over the water with outstretched neck, their beating wings producing a humming or whistling sound. Look for white patches on wings.

Visitors from Iceland supplement our resident

population in winter months. It is estimated that 10,000 pairs breed in Finland, but the size of the large Soviet Union population is not known. About 20,000 birds spend mid-winter in the Baltic. The total population for north-western Europe is put at 40,000 individuals.

For winter-watching, the Cromarty and Beauly Firths, Highland; Poole harbour, Dorset; Traeth Bach, Gwynedd; and Dyfi estuary, Dyfed, are some of the better areas.

The nominate *M.s. serrator* breeds in N. Europe, Asia, and N. America, and the sub-species *M.s. schioleri* in Greenland; the latter is a little larger with a stronger and broader bill.

Goosander (*Mergus merganser*)
Length male 67 cm (26¼ in), female 63 cm (25 in)
Fig. 11 (1 and 1a)

The British population is thought to be a little over 1,000 pairs; these are mainly in Scotland with small numbers in Northumbria and the Lake District. The goosander breeds inland close to lochs and rivers, in or near to wooded areas; they nest in a natural tree cavity, and the eggs are laid from April to June. Even in winter it much prefers freshwater lakes and rivers, and is far less likely to be seen in British coastal waters than the red-breasted merganser (*M. serrator*). Even so, in winter when the Arctic and sub-Arctic lakes become frozen-over, large numbers leave for the coast in search of food. For instance up to 30,000 have been recorded along the Danish coastline, 10,000 in the Mediterranean-Black Sea area. The total European population is put at 75,000 birds.

The drake is easily distinguished from the male red-breasted merganser, the only real similarities being the bottle-green head (but no crest) and long slender hooked bill. His breast and sides are creamy-white. Mantle glossy black; centre of back grey, shading darker towards tail. Wings black with white coverts, and all but the innermost secondaries are white narrowly edged black on outer web. Female has head (with drooping crest) and neck of rich chestnut brown; this is sharply defined from the otherwise grey upperparts. Belly pale. Both sexes have a long red serrated bill, also reddish-orange legs and feet.

The diet consists almost entirely of fish. Both sexes utter a low quack-like call in winter.

The nominate *M.m. merganser* breeds Iceland and Europe to N.E. Asia; *M.m. orientalis*, C. Asia to Himalayas; and *M.m. americanus*, Canada and W. USA. One good place for winter-watching is Beauly Firth, Highland.

White-fronted Goose (*Anser albifrons*)
Length 70 cm (27½ in)

Although this is Europe's most numerous goose with a total population in excess of 300,000, only small numbers (from two of its breeding populations) winter in the British Isles. It is slightly smaller than the grey-lag goose (*A. anser*), but with similar plumage. Look for black bars across chest and belly; perhaps more readily identified by the white forehead (hence its name). The bill is some shade of pink (not orange-coloured); legs and feet are mostly orange (but occasionally pale pink as the grey-lag's).

Of the five recognised sub-species, only the Greenland white-front (*A.a. flavirostris*) (which is darker and larger than the nominate), and the Russian or European white-front (*A.a. albifrons*) (nominate), winter in the British Isles.

Of our wintering white-fronts, 16,000 are the Greenland race; half of which are to be found in an area of reclaimed marshland, the Slobs; this extends on either side of Wexford harbour in southern Ireland. A further 3,000 visit Islay in the Outer Hebrides. Of the Russian or European race *A.a. albifrons* only 7,000 to 8,000 birds winter with us; these numbers are very dependent on the severity of the weather in its other European wintering haunts. They are mainly found at New Grounds, Slimbridge, on the River Severn; and the Towy Valley, South Wales.

The staple diet of the European white-front seems to be almost exclusively grass of one species or another. In Ireland the Greenland white-fronts feed on the roots of cotton grass (*Eriophorum*) and the bulbils of the beak-sedge (*Rhynchospora*), which they dig up with their long bills.

The other three sub-species are the Pacific white-front, *A.a. frontalis*, of Eastern Siberia and Northern Canada; *A.a. gamnelli* of North Western Canada; and *A.a. elgasi* of Alaska.

Other places to watch the white-front are the Swale in Kent and at Thames-side marsh on the north Kent coast.

Flighting pink-footed geese (*Anser brachyrhynchus*).

Some coastal areas for winter-watching are the south bank of the Solway, Cumbria; Ribble estuary, Lancashire; Ythan estuary and neighbourhood, Grampion; north bank of Solway, Dumfries and Galloway; Tay estuary, Tayside; and the Wash, Lincolnshire.

Grey-lag Goose (*Anser anser*)
Length 80 cm (31½ in)

This is the largest of the 'grey geese', and the only goose indigenous to Britain. The wild population of grey-lags has been almost eliminated during the past two centuries, therefore most of the British stock is comprised of feral birds. Truly wild populations can be found in the Scottish Highlands, and the Outer Hebrides.

In general appearance a greyish-brown goose with lightly barred upperparts, the feathers having pale buffish margins so producing this effect. Underparts similarly marked but shading to ashy-grey on lower breast and to white on belly. It has a large orange bill; pink legs and feet. Sexes similar. Call a three-noted 'honk' typical of farmyard geese.

In Scotland the breeding season commences during late April. The nest is a large depression amongst heather, or an elevated site in a marshy region, with a lining of down and feathers. The clutch size varies a great deal and may be from three to twelve creamy white eggs; four to six would be a normal clutch.

The female alone undertakes the incubation which lasts for 27 to 28 days; she leads her goslings away to a suitable feeding area when only a few hours old, the male bird bringing up the rear. Young grey-lags fledge in about eight weeks.

The food includes roots, grasses, and the rhizomes of marsh plants.

Almost the entire Icelandic population of over 60,000 birds arrive in Britain to spend the winter. These birds have a liking for corn stubble, and for growing root-crops including swedes and carrots.

There are two races, *A.a. anser* (the nominate) of N. Europe and N. Asia, and *A.a. rubirostris* of C. and E. Asia.

For winter-watching at the coast: the Ythan estuary and nearby areas, Grampion; the Beauly and Inner Moray Firths; Tay estuary, Tayside; and Lindisfarne (Holy Island), Northumbria.

Pink-footed Goose (*Anser brachyrhynchus*)
Length 60–70 cm (23½–27½ in)

This is often regarded as a sub-species of the bean goose (*Anser fabilis*), but here for reasons of convenience we shall treat it as a separate species. There are two distinct breeding populations. the Spitsbergen (of 12,000 to 15,000 birds) whose main wintering grounds are in the Netherlands; and those from Eastern Greenland and Iceland (together totalling around 70,000 birds). The first arrival of wintering birds occurs in early September with peak numbers arriving during October.

The adult bird has feathers of mantle, wings, and sides of body greyish-brown with prominent pale buff or white tips, thus producing a somewhat barred appearance. Breast and belly are pale brown, narrowly and far less distinctly barred. Head a uniform brown. Bill black with a pink band towards tip. Legs and feet pink (the bean goose is generally darker overall, and has an orange-yellow bill, which is black at tip and towards base of upper mandible; legs orange-yellow). Sexes similar.

On arrival in September/October it feeds on corn stubble and spilled grain. It later feeds on growing root-crops, especially potatoes and carrots (particularly when affected by frost).

The call is a far-carrying 'wink-wink' which usually indicates the presence of a skein flying high above in that typical straggly V-formation, or perhaps as an extended line.

Brent Goose (*Branta bernicla*)
Length 55–60 cm (21½–23½ in)

The very dark plumage of this small goose makes identification relatively easy. Adult birds have head, neck, and upper breast completely black, apart from small patches of white on upper neck; these patches form a partial or sometimes an almost complete white collar. Upperparts dark greyish-brown; upper and under tail-coverts, and sides of rump, are white (these areas of white produce a V-shape when bird is in flight). The brent goose's underparts vary from pale brownish-grey to almost black, depending on the sub-species. Bill, legs and feet are black. Sexes similar.

Very much a breeding bird of the high Arctic, with circumpolar distribution, which only visits our shores in winter. Its staple diet is the eel grass (*Zostera*), which grows between high and low water mark in shallow muddy coastal areas. Voice a guttural 'krronk'.

Of the four sub-species, the dark-bellied brent, *B.b. bernicla* (nominate), from western Siberia, winters in Denmark, the Netherlands, France and England (of the 140,000 birds that arrive on the southeast coast, most will stay there, but small numbers will disperse to the Norfolk and Suffolk coasts, and other areas). The light-bellied or Atlantic brent, *B.b. hrota*, of the Greenland and north Canadian population, arrives on the northern shores of Ireland and disperses around the coast (these may number 6,000 to 12,000). A few light-bellied brents of the Spitsbergen population will winter on the Northumbrian coast at Lindisfarne (Holy Island). The other two sub-species are *B.b. nigricans* (the black or Pacific brent), which breeds in the high Arctic regions of N.E. Canada; and *B.b. orientalis* of E. Siberia and N.W. Canada.

Some good areas for winter-watching are as follows. *Dark-bellied*: the Wash, Lincolnshire; Foulness and Maplin sands, Essex; and Langstone harbour, Hampshire. *Light-bellied*: Strangford Lough, Ireland; and Lindisfarne, (Holy Island), Northumbria.

Barnacle Goose (*Branta leucopsis*)
Length 60–70 cm (23½–27½ in)

There are three distinct populations of barnacle geese, one in eastern Greenland, another in Spitsbergen, and a third on the Novaya Zemlya islands between the Barents and Kara Seas north of Siberia. The Greenland birds winter in Ireland and western Scotland, the most important wintering area seeming now to be on Islay in the Inner Hebrides, where numbers of over 20,000 have been recorded. Those breeding in Spitsbergen winter on the Solway Firth, and number about 8,000. The third group from Novaya Zemlya travel south to winter in the Netherlands and are in order of 40,000 to 50,000 birds. The world population is put at about 80,000.

Barnacle geese are quite easy to identify, with face and forehead creamy-white, and a narrow black band extending from front of eye to base of bill. Crown, neck, and breast are glossy black. Upperparts dark grey, the feathers have a black subterminal band and are tipped with white, thus producing a heavily-barred effect. Upper and under tail coverts are white; tail black. Underparts white but flanks are faintly barred with pale grey. Bill short and black; legs and feet black. Sexes alike.

The voice is a short shrill bark, repeated several times in quick succession. In winter feeds principally on the stolons (horizontal shoots which run along the ground surface) of white clover (*Trifolium repens*); this grows on 'merseland' (which is seawashed pasture), along with salt-marsh rush (*Juncus gerardii*) and fescue grass (*Festuca rubra*).

Where to watch in winter: north bank of the Solway in an area centred on Caerlaverock and the Wildfowl Trust Refuge (Spitsbergen population). The island of Islay, Inner Hebrides (birds of the Greenland population); also south bank of the Solway (Spitsbergen population).

Bewick's Swan (*Cygnus bewickii*)
Length 120 cm (47 in)

The smallest of the three white European swans. Look for bill which is black to base; the yellow area on each side ends abruptly behind the nostrils. The black and yellow patterning of bill is so variable that individuals can be recognised by this characteristic alone; a fact first discovered by Sir Peter Scott. Breeding is restricted almost entirely to the tundra regions of northern Russia and Siberia, and extends from the Kola peninsula and Petsamo to the Lena delta. Jankowski's swan (a sub-species of the whistling swan) replaces it eastwards beyond the River Lena.

About 200,000 to 300,000 Bewick's swans winter

in the British Isles. Among the few regular sites are the Ouse Washes, Cambridgeshire, and the River Severn, Gloucestershire. Ireland has a more widely-scattered number of wintering areas containing far fewer birds. The majority arrive on our shores during January and February.

Voice resembles the bugling of a whooper swan (*C. cygnus*) but quieter. The alarm note is an oft-repeated harsh 'howk'.

Food mainly vegetable matter, with the eel grass (*Zostera*) its staple diet when feeding coastally.

Nowadays often considered as a sub-species of *Cygnus columbianus*, the Whistling Swan of North America. Hence *C.c. columbianus*, whistling swan, N. Canada; *C.c. bewickii*, Bewick's swan, N. Russia and N. Siberia; and *C.c. jankowskii*, Jankowski's swan, N.E. Asia.

For winter-watching at the coast, Ribble estuary, Lancashire; the Wash, Lincolnshire; north bank of the Solway, Dumfries and Galloway; Wildfowl Trust and surrounding area, Gloucestershire; Lough Neagh, N. Ireland; also Wexford Slobs, southern Ireland.

Whooper Swan (*Cygnus cygnus*)
Length 150 cm (59 in)

The difference between the whooper swan and the Bewick's swan (*C. bewickii*, length 120 cm, 47 in), is mainly one of size. At close range look also at the bill pattern, which is yellow from the base and along each side narrowly almost to tip, the rest black. The plumage is entirely white; legs and feet are black. Being a wild bird it is much more difficult to approach than the mute swan (*C. olor*).

Its breeding range extends across N. Europe (including Iceland) and N. Asia. Birds wintering in Britain are probably of the Icelandic population *C.c. islandicus*. They begin to arrive in large flocks during late September; as winter approaches so they disperse as small groups (usually of less than 50 birds). Mostly encountered on shallow lochs and lakes throughout central and southern Scotland, northern England, and Ireland. Of the estimated 6,500 whooper swans visiting the British Isles, about 2,000 are in Ireland.

Coastal areas are also visited where the eel grass (*Zostera*) is the staple food. Call a double-noted bugling, the second note of higher pitch.

The nominate *C.c. cygnus* breeds in N. Europe and N. Asia; *C.c. islandicus*, S. Greenland and Iceland; and *C.c. buccinator*, the trumpeter swan (which is often regarded as a separate species) of N. Canada and N. USA.

For coastal watching in winter, the following are worth visiting: Lindisfarne (Holy Island) in Northumbria; Ythan estuary and surrounding areas in Grampian; Cromarty Firth in Highland; and the Ribble estuary in Lancashire.

Mute Swan (*Cygnus olor*)
Length male (cob) 150 cm (59 in), female (pen) smaller.

The mute swan's orange bill with black basal knob is sufficient to distinguish it from the other two white European swans. Look also at the necks; the mute swan's is usually held in a gentle curve, whilst the Bewick's (*C. bewickii*) and the whooper (*C. cygnus*) hold their necks straight for most of the time. In Britain the mutes are of semi-domesticated origin; only occasionally do truly wild birds occur in winter. About 20,000 pairs breed in Europe, at least half of which are in Scandinavia. Britain has probably less than 5,000 breeding pairs, the greater number of which are in Ireland; there is possibly an equal number of non-breeders.

Not strictly voiceless as the name would suggest, it produces various grunting and hissing sounds.

Breeding usually commences in late April or early May. A huge pile of materials is accumulated to form a nest, including reed stems, rushes, roots, leaves, and sticks. Both birds help to gather it, but the female makes the final arrangement.

The greyish-white eggs usually number four to seven, they are incubated for about five weeks. Often the parents carry the cygnets aboard their back, sheltering them between half-open wings.

Food is almost entirely submerged vegetable matter, such as stems and roots of aquatic plants; tadpoles, small frogs and insects are also taken. When feeding in coastal areas the eel grass (*Zostera*) is a favourite food.

The mute swan breeds from Europe to C. Asia; there are no sub-species.

Places to watch in winter include: the Fleet, Chesil Beach, Dorset; Chichester harbour, Sussex; Inner Thames, Kent; Stour estuary, Essex; and Cromarty Firth, Highland.

Avocet (*Recurvirostra avosetta*)

Length 43 cm (17 in)

During 1947, after a non-breeding period of 100 years, a few pairs began to breed at Havergate and Minsmere in Suffolk. Havergate now has about 100 breeding pairs, and Minsmere over 50 pairs. So successful has the return to Britain been that today nesting also occurs in Essex and Kent.

The avocet is a graceful long-legged bird with a long, slender, and markedly upturned bill. Its plumage is mainly white but with a black crown and nape; there are two converging black bands on its back, two black wing bars, and the wing-tips are also black. In flight the long blue-grey legs extend beyond the tail. Bill black. Sexes alike.

Usually a colonial nester, although solitary nesting is not unknown. The nest is usually just a scrape on hard sunbaked mud, or a depression in a grassy situation. The four eggs are laid in mid-April, the ground colour is some shade of olive green, boldly blotched with blackish-brown. Both sexes share the incubation for 22 to 24 days, and the young fledge after about six weeks. If the nest site is approached the avocet utters a continuous screaming 'kleet-kleet-kleet-kleet'. In flight a loud and high-pitched 'klooit-klooit' is uttered repeatedly.

The avocet has an unusual method of acquiring food; whilst wading in water it sweeps the surface from side to side, skimming off small invertebrates with its partly-opened bill.

Its distribution as a breeding bird extends from Europe to China; it also occurs in India and S. Africa. There are no sub-species.

During autumn the breeding territories are vacated, and coastal areas in south-west England provide winter quarters for a hundred or more, especially the Tamar estuary in Devon, and to a lesser extent the Exe estuary.

Goldeneye (*Bucephala clangula*)

Length drake 47 cm (18½ in), duck 43 cm (17 in)

Small numbers of goldeneye are known to have bred in Scotland since 1970. Male has dark green head, the prominent roundish white patch in front of eye is the best means of identification. (The white cheek marking of Barrow's goldeneye, *B. islandica*, is more of a crescent-shape.)

The body is mainly white, the back black. Wings are black, striped with white (wings of Barrow's goldeneye are black with white 'port-holes', not stripes). It has an almost black bill; orange-yellow legs and feet with dusky webs. The female's head is dark chocolate brown; most of her neck, and from lower breast to vent, are white. Upperparts are mottled with slate-grey and blackish-brown. Bill blackish with a narrow dusky orange band near the black nail. Legs and feet are yellow with dusky webs.

In eclipse plumage the males are similar to the females but a little brighter; bill black. Immature males are similar to the adult males in eclipse.

Mostly a silent bird except during the display when the drake utters a rattling call and the female a gruff croak. The drake's display includes 'head flicking' from side to side; followed by a 'head throw' when the head is thrust forward and then lowered backwards onto the rump.

When breeding, it frequents the vegetative margins of lakes, pools, shallow rivers, and slow-running streams, especially those close to mature woodland (the forest tundra zone of Russia and N. Scandinavia).

A natural tree cavity or the disused nest-hole of a large woodpecker is the preferred nest site, but nest-boxes are readily used if available. Laying commences during early May in its more southerly distribution, and up to a month later further north. An average clutch would be nine to ten bluish-green eggs. Incubation is by the female alone and takes about 28 days; the drake stays in the vicinity for at least part of this period.

During summer months, caddis-fly and midge larvae are taken from pools within its woodland habitat; also frogs, tadpoles and freshwater mussels. The principal food in winter is mussels, shrimps, and small crabs.

There are an estimated 170,000 pairs breeding in Finland and western Russia; and a wintering population of 150,000 birds in north-western Europe (two-thirds of which occur in the Baltic). The Baltic also holds several thousands of moulting goldeneye in summer.

Some good places to watch in winter are Leith to Musselburgh, Lothians; Cromarty Firth, Highland; Chichester harbour, Sussex; and the Wash, Lincolnshire.

The nominate *B.c. clangula* breeds in N. Europe and N. Asia; the American sub-species *B.c. americana* breeds in Canada.

Common Sandpiper (*Tringa hypoleucos*)
Length 19.5 cm (7½ in)

Upperparts, including neck and wings, are brown; breast and underparts pure white; neck faintly streaked with brown. On the ground it is easily identified by its persistent bobbing up and down, both when standing or running about. During summer months it frequents lakes, moorland reservoirs, hill streams, and other freshwater areas. After breeding, coastal lagoons and estuaries are amongst its feeding haunts; seldom encountered on open sand or mudflats. Not what we would class as a true coastal species. After breeding, small parties on their southward passage may be seen along the coast.

As a breeding species its distribution includes most of the northern hemisphere.

Ruff (*Philomachus pugnax*)
Length male (ruff) 29 cm (11½ in), female (reeve) 23 cm (9 in)

During winter months both sexes become boldly scaled with blackish-brown and sand colour on upperparts, and buffish-brown below. At this time

Avocet (*Recurvirostra avosetta*).

the male bird does not have that familiar ruff or eartufts. Bill shortish relative to body size.

As with the common sandpiper (*Tringa hypoleucos*), the ruff is not really a coastal species even in winter, but is occasionally encountered in estuaries or marshy coastal areas.

Divers (Gaviidae)
Fig. 12

The black-throated diver *Gavia arcticus*, length 56–68 cm (22–26½ in), body 36–43 cm (14–17 in), is larger than the much more common red-throated diver, *G. stellata*, length 56–69 cm (22–27 in), body 36 cm (14 in); but smaller than the great northern diver, *G. immer*, length 70–90 cm (27½–35½ in), body 43–50 cm (17–19½ in) which is a very rare breeding species in the British Isles, and primarily a North American breeder.

Fig. 11 1. Red-throated diver (*Gavia stellata*). 2. Black-throated diver (*Gavia arcticus*). 3. Great Northern diver (*Gavia immer*). Birds in winter plumage.

All three species spend the winter at sea. Even the Great Northern can become quite common in the coastal water of Britain and Ireland, probably birds which have bred in Greenland or Iceland; generally speaking, it has a stouter bill and a thicker neck than the other two species.

When winter-watching for divers in coastal waters look for the following features. *Red-throated diver*: pale face, speckled back, and thin up-turned bill. *Black-throated diver*: dark back, paler crown than great northern diver, slender bill not upturned. *Great northern diver*: dark back and crown, bill straight and stout, neck thick.

Grebes (Podicipitidae)
Fig. 13

The great crested grebe (*Podiceps cristata*), length 48 cm (19 in), body 30 cm (12 in), is our largest grebe and nests commonly in the British Isles. The Slavonian grebe (*P. auritus*) and the black-necked grebe (*P. nigricollis*) are both a little smaller, length 30–33 cm (12–13 in), body 18–20 cm (7–8 in), and both breed sparingly in the British Isles; the latter only three or four pairs in most years. The fourth species likely to be seen in coastal waters during winter is the red-necked grebe (*P. grisegena*) which does not breed in Britain, length 43 cm (17 in), body 25 cm (10 in).

When winter-watching for coastal-swimming grebes it is worth bearing in mind that the great crested is the one most frequently seen. It looks very whitish, the white extending above eye, the bill is pinkish. It is most easily confused with the red-necked grebe which is of similar size but has a grey neck, no white above eye, and has yellow base to bill. It is also most likely to be seen in the coastal waters of east and south-east England. The Slavonian grebe is much smaller than the two previously mentioned, and in winter has white on sides of face, front of neck, and upper breast; back dark. Bill straight. It can be seen on southern coasts of England and Wales, also off the west and east (less generally) coasts of Scotland, including coastal waters of Orkney and Shetland, also in Irish waters. It seems to favour sheltered bays and estuaries. The black-necked grebe, which is of similar size to the Slavonian, has front of neck grey, upturned bill, and black cap which extends below the eye, back darkish. Most likely to be seen off the southern coasts of England and Wales. It is almost exclusively a wintering migrant.

Sanderling (*Calidris alba*)
Length 20.5 cm (8 in) **Fig. 14(3)**

The main breeding areas of this charming little wader are in eastern Greenland, parts of northern Siberia including its off-shore islands, north-west Spitsbergen, Alaska, and parts of Arctic Canada.

During the summer season its upper plumage is chestnut with black speckles, many of the feathers edged with white; the underparts are pure white. Bill and legs are black.

When the sanderling reaches our shores on its winter migration from the now frozen high Arctic, its plumage is much more sombre. The head and most of the underparts are white, the upperparts have lost their russet suffusion, and are now pale grey with a few slightly dark markings. Look for the characteristic black shoulder-patch. Of all the small waders that visit Britain during winter, the sanderling is by far the palest.

As it feeds in 'small' groups running along the water's edge it is quite a confiding little bird; not so when it occurs in 'large' numbers. It may be seen in the company of other small waders such as the dunlin (*Calidris alpina*) which is slightly smaller and a little darker.

Morecambe Bay, Ribble and Dee estuaries, south bank of the Solway, the Wash, Teesmouth, Humber estuary, and Chichester harbour, are amongst the best places for watching in winter.

Purple Sandpiper (*Calidris maritima*)
Length 21 cm ($8\frac{1}{4}$ in) **Fig. 7(2)**

Of all small wading birds to visit our shores in winter the purple sandpiper is the most confiding, and a close approach is possible – with care. Like so many of the waders that breed in the high Arctic its summer plumage is a mixture of dark browns suffused with rufous. We in Britain see them in a less colourful attire. However, it is darker than most other of our small waders, with head, breast, and upperparts deep brown and with a hint of purple; underparts white. It is often seen in small parties quietly feeding at rock-pools left by the ebb-tide. Small crustaceans and molluscs are sought for under seaweeds or picked out from rock crevices.

Fig. 12 1. Red-necked grebe (*Podiceps grisegena*). 2. Great crested grebe (*Podiceps cristata*). 3. Black-necked grebe (*Podiceps nigricollis*). 4. Slavonian grebe (*Podiceps auritus*). Birds in winter plumage.

Fig. 13 1. Grey plover (*Pluvialis squatarola*). 2. Curlew sandpiper (*Calidris ferruginea*). 3. Sanderling (*Calidris alba*). 4. Little stint (*Calidris minuta*). Birds in winter plumage.

When in flight, as a small flock, an almost swallow-like twittering may be heard.

Seldom seen in groups exceeding 30 or 40 birds. The dark sombre winter plumage often makes it difficult to spot against its surroundings.

The purple sandpiper is given the distinction of being probably the most northerly wintering wader in the entire world.

The Firth of Forth and the Dee estuary are two of the best areas to watch for this species in winter.

Curlew Sandpiper (*Calidris ferruginea*)
Length 19 cm (7½ in) **Fig. 14(2)**

In northern Europe is only to be seen along the shores when on passage. At such time the upperparts are mainly streaked with brownish-grey, and white beneath; closely resembling the dunlin (*Calidris alpina*) with which it freely associates. On the ground its downward curved bill may be inadequate to separate it from the dunlin (which also has a slightly decurved bill). Look for the longer legs and neck, smaller head with a more pronounced pale superciliary (eyebrow) and dark eye-stripe.

In flight the white rump and blackish tail become a more apparent means of differentiation. When at rest the blackish wing-tips tend to obliterate the white rump.

Unfortunately we seldom see this bird in its full breeding plumage. Then it is predominantly brick red, patterned in blackish-brown. The bright rufous of underparts, neck and sides of head are mottled and barred.

The only definite breeding areas of the curlew sandpiper appear to be in northern Siberia, extending eastwards from the Yenesei delta to 175°E and north to 74°N, including the Taimyr Peninsula, the Liakhof Isles and others.

Whilst the principal wintering quarters are in the Mediterranean, south-east Asia, Africa as far south as South Africa, and Australasia, it is a regular passage migrant to Western Europe, including Britain (but in small numbers).

In North and South America it occurs only as rare vagrant. Even though its breeding range is so restricted, it truly spans the world as a migrant species.

The Tay estuary and Chichester Harbour are particularly good areas for winter sightings.

Common Scoter (*Melanitta nigra*)
Length 47 cm (18½ in)

The only species of duck with completely black plumage (male); the female is predominantly chocolate brown. In its wintering quarters, the coastal waters around our shores, it could only be confused with the velvet scoter (*Melanitta fusca*) which does not breed in the British Isles. The latter on *close* inspection has a small whitish crescent beneath the eye and an area of white on the secondary wing feathers; this white wing patch is unmistakable in flight. The common scoter has a black knob at base of bill; there is a patch of yellow between this and the bill tip. The velvet scoter's bill, however, appears mainly orange, a deeper shade towards the tip, with a black basal knob. At 56 cm (22 in) it is a larger bird.

Shows a preference for the less sheltered coastal waters, and during winter months can be seen in scattered groups on open water, or as individuals strung out and swimming at speed, or diving for mussels and periwinkles (their principal food in winter). The most vocal of our scoters, at this time of year the piping call of the males can be heard from flocks out at sea.

As a breeding bird its range extends from N. Ireland across northern Scotland (including the Orkneys and Shetlands) to northern Scandinavia and most of Siberia (on either side of the Arctic circle); also Iceland and the Faeroe Islands. Full clutches of eight or nine creamy or buff-coloured eggs are to be found from late May onwards, but peak laying occurs during the first half of June. The nest-site is usually at the edge of a loch or large lake, often on an island therein, and preferably in a dry area amongst grasses or sedges under cover of dwarf birch or willow scrub. Incubation is undertaken by the females alone and for a period of 27 to 31 days. Now that their mates are thus preoccupied the males leave the breeding grounds and fly out to sea where they gather in large rafts in traditional moulting areas off the coasts of Britain, Holland, and especially Denmark where the numbers of moulting males often reach 200,000.

The wintering population of common scoters in European waters is estimated at about 450,000 birds. Some important wintering sites in the British Isles are in the Lothian Region of Scotland: Leith–Musselburgh; Aberlady Bay and Gullane Bay. In recent years large flocks of up to 25,000 birds have been seen at Carmarthen Bay in South Wales. Common scoters are very vulnerable to oil pollution when spillage occurs in coastal waters.

The other recognised race is the American scoter, *M.n. americana*; this has a larger area of yellow on the bill but a smaller black basal knob. It breeds in eastern Siberia, western Alaska, and locally in British Columbia (Canada).

Canada Goose (*Branta canadensis*)
Length 60–110 cm (24–43 in), depending on race

It is a remarkable fact that both the smallest and the largest of all true geese are found in this one species. All have the same basic plumage pattern, but some races are darker than others. The British population is thought to be in the region of 20,000 birds, and these are found mainly in the Midlands and southern counties. They are seen mostly on meres, reservoirs, and park lakes; they are not wild stock, but have been distributed and transported by man from area to area. A meeting with truly wild birds breeding on the Canadian 'muskegs' is an experience we shall long remember. Only very occasionally does a truly wild bird turn up in Europe.

An adult Canada goose has crown, front of head, and neck black; with a conspicuous white cheek patch which may or may not be continuous under the throat. In some races a whitish area may extend round base of neck. There is no furrowing of neck feathers. Body plumage varies from greyish-brown to dark brown, the feathers having buff edges, thus producing a barred effect. Underparts from belly to tail are whitish, and there is a crescent of white on rump. Bill, legs and feet black. Sexes similar. The large races utter a loud resonant 'honk'. Food is mainly vegetable, including various sedges, grasses and aquatic plants.

There are ten races of which the nominate, *B.c. canadensis*, has been introduced into Europe.

In winter, Canada geese can be seen at Poole harbour, Dorset; and Newton marsh, Isle of Wight. These are not wild birds.

Smew (*Mergus albellus*)
Length 40 cm (15½ in) **Fig. 11(3 and 3a)**

By far the smallest of the three European 'sawbills'; it does not breed in Britain but winters in south-east England and is not confined to coastal areas. Prefers inland waters when free of ice. There are important wintering areas in the Netherlands with over 10,000 birds; the coasts of the Black Sea and the Sea of Azov together hold about 65,000. It is thought that the European population may be on the increase.

The mature drake is a mainly white bird with small areas of black, such as the eye patch which extends to base of the bill; and another to rear of crest, the back has black markings which continue as two narrow bands on either side and then extend down middle and lower breast respectively. The rump and wings have areas of greyish-black. Legs and feet are dark greenish-grey. Bill dark grey and not obviously serrated. Female has crown (down to eyes), and back of neck chestnut-brown. Lower part of head and throat are white. Sides and upperparts are mottled medium grey; belly white. Bill and legs similar to the drake's.

When winter-watching at the coast perhaps most likely to be seen from the Wash southwards, and then westwards to about Portland Bill.

As a breeding species occurs in N. Europe and N. Asia; there are no sub-species.

Long-tailed Duck (*Clangula hyemalis*)
Length 53 cm (21 in) **Fig. 15**

It is said that, on average, the long-tailed duck breeds further north than any other species of duck. The tundra and tundra-forest regions throughout the Arctic are its summer haunts. The European population is estimated at 500,000 birds, with about 700 pairs breeding at Lake Myvatn in Iceland; and there are said to be 740,000 pairs breeding in western Siberia. The Baltic holds most of Europe's 100,000 wintering birds.

The drake has the distinction of having three different plumage phases annually; here we shall describe its winter dress (November to April). This is the most handsome, for it is during that period when courtship takes place. The drake is mainly black and white; the white crown has a central area of yellowish-orange; nape, back of neck, and upper breast are white; sides of head pale grey with a white ring around the eyes; on the rear of its cheek there is a large sooty-black patch. The lower breast, backs of wings, and tail are blackish-brown (the tail streamer very long, sometimes up to 24 cm). Sides of body appear pearly-grey, underparts are white. Bill slatey-blue with a broad pinkish band and a blackish

Fig. 14 Long-tailed ducks (*Clangula hyemalis*), male (above) and female. Birds in breeding plumage.

nail. Legs and feet are pale bluish-grey; webs dusky. The female at this time has head, neck, and back dark chocolate brown (feathers on back have chestnut borders). There is an oval area of pale grey around the eyes; this extends backwards as a line down the neck. The breast and sides are barred and mottled pale brown and whitish. She has no tail streamers. Bill slatey-blue.

About 90 per cent of its food intake is animal matter including crustaceans, and a variety of molluscs, also insects and fish. Seeds of pondweeds and grasses make up the remaining 10 per cent.

The long-tailed duck's most common call is a loud resonant 'ow-owooolee', it is heard mostly at sea.

Perhaps the best places to watch in winter are, Aberlady Bay, and Leith–Musselburgh, Lothians; Lindisfarne (Holy Island); the Wash, Lincolnshire; also Ythan estuary and neighbourhood, Grampian.

It breeds in arctic regions of N. Europe, Asia, and North America. There are no sub-species throughout this range.

Scaup (*Aythya marila*)
Length 48 cm (19 in)

Outside the breeding season the scaup spends most of its life in shallow coastal waters and estuaries. For instance the Baltic may hold winter flocks of up to 80,000 birds. Icelandic breeders concentrate in the Firth of Forth where flocks of up to 20,000 have been recorded. There are, however, large winter flocks showing favour for freshwater; one example is the Ijsselmeer in the Netherlands where numbers of up to 25,000 have been recorded. Upwards of 30,000 often gather in the Waddenzee. This brings the total number of wintering scaup in north-western Europe to 150,000 birds.

The male in breeding plumage has neck, breast, upper back, and rump black (the feathers on head having a greenish gloss). Upperparts are pale grey with black vermiculations. Belly, flank, and sides are white. Bill pale blue with a black nail. Legs and feet greyish-blue, webs dusky. The female's head, front of neck, and breast are dull dark chestnut; there is a conspicuous white patch at base of pale blue bill. Upperparts are dark brown flecked with buff, underparts much lighter due to pale edging on feathers. Belly whitish. Usually a silent bird outside the breeding season.

The scaup feeds mainly on mussels, cockles and periwinkles, which it acquires by diving; but it is not entirely carnivorous and when feeding on inland waters quantities of seeds, pondweeds and sedges are also eaten. The wintering flock on Islay depends to a large extent on waste grain from a local distillery.

The nominate *A.m. marila* breeds in both open and scrubby tundra, and in coastal areas void of trees. Its range extends across the Arctic from N. Europe to E. Siberia, including Iceland (which has about 5,000 pairs).

Places for winter sightings: Leith to Musselburgh (Firth of Forth); Islay, Inner Hebrides; and Poole harbour, Dorset.

The North American, Pacific scaup, *A.m. nearctica*, is one of two sub-species, and breeds mainly in Alaska, also N.W. Canada; the other is *A.m. mariloides* of the Bering Is.

Golden Plover (*Pluvialis apricaria*)
Length 28 cm (11 in)

In summer dress the upperparts are black, spotted with rich yellowish-gold; the underparts from the black face and throat down to lower belly are jet black. The white across forehead and above eye continues down sides of the black face, throat and belly. Bill black. Legs and feet dull grey. During winter the gold of upperparts appears less rich, face and underparts become whitish, and breast is mottled brown.

When in breeding plumage the British population, which numbers between 25,000 and 30,000 pairs (but slowly decreasing), can be distinguished from the northerly breeders by the broader and more distinct white border of their black underparts. Of this population about ten pairs breed in the south-west, and about 500 pairs in Wales. Large numbers are not found south of the Northern Pennines.

For breeding, a moorland habitat is preferred. Egg-laying occurs between April and June, the nest is a shallow depression in the ground, sparsely lined with grass; the four yellowish-buff eggs are heavily and boldly blotched with blackish-brown, especially at the broad end. Incubation, which is shared by both sexes, continues for about 30 days.

In winter months severe weather often causes them to leave their lowland feeding areas and move to mudflats along the coast. The north and south banks of the Solway; Camel estuary, Cornwall; Torridge-Taw estuary, Devon; Hamford water, Essex; Morecambe Bay and the Ribble estuary are some of the places worth watching in winter.

There are two forms: *P.a. apricaria* (the nominate) of N. Europe and N. Asia; and *P.a. oreophilis*, N. and W. British Isles, Denmark and Germany.

Grey Plover (*Pluvialis squatarola*)
Length 28 cm (11 in) **Fig. 14(1)**

This is the largest of the North American plovers. The male in summer dress is truly magnificent; the white upperparts are marked and spotted brownish; the forehead and sides of neck are white; thus contrasting sharply with the black face, throat and upper breast; belly white. The feathers of the female's upperparts are tipped white, and her underparts not of such a solid black as her mate's and flecked with white.

The grey plover breeds in a narrow band that extends along the coast of northern Russia eastwards from Kanin and Kolguev to the Bering Straits; then to Alaska; Baffin Island and Southampton Island (Canada). There are no sub-species throughout its range.

When they reach our shores in winter, the face, sides of neck, throat, breast and belly are no longer black. The bird is more or less entirely pale greyish, but is marked and spotted a brownish colour.

The total British wintering population seldom exceeds 15,000 birds, most of which occur in the southern counties; but the Wash holds the highest counts.

Areas to watch in winter are: the Wash; Ribble estuary; the Swale, Kent; Medway marshes and estuary, Kent; Hamford water, and Blackwater estuary, Essex; and Chichester harbour, Sussex.

Little Stint (*Calidris minuta*)
Length 13.5 cm (5 in) **Fig. 14(4)**

One of Europe's smallest waders, about the size of a house sparrow. An Arctic breeder of northernmost Scandinavia, but principally of Russia and Siberia. Like so many of the Arctic breeding waders its summer plumage is predominantly reddish brown.

In winter months, when seen on British coasts the plumage has become much more sombre; with upper parts mottled buff and black, giving a fawny-grey appearance; the underparts are white suffused with buff. Look for a pale V-shape on upper plumage which points backwards towards the tail. Has a narrow white wingbar, and dark centres to rump and tail. Bill and legs are black.

For winter-watching: Poole harbour, Dorset; Pagham and Chichester harbours, Sussex; Lindisfarne (Holy Island); and the Tay estuary, Scotland.

List of Plant Species

Sea Plantain (*Plantago maritima*)
Family Plantaginaceae (112) **Fig. 16(4)**

A common and usually glabrous perennial which is more or less confined to maritime conditions, but can occasionally be found growing alongside mountain streams in the Snowdonia range, the Pennines, and northern Scotland. Its narrow, linear radical leaves have three to five indistinct veins, are fleshy, entire or slightly toothed, usually 5–15 cm (2–6 in) long, and up to 15 mm broad; but others are much smaller and almost grass-like. The tiny brown and yellow flowers vary in number and are borne on a smooth, rounded scape (the flowering stem of a plant whose leaves are all radical) of up to 30 cm (12 in) tall which is usually longer than the leaves, or at least of equal length. Flowerspike usually 2–6 cm ($\frac{3}{4}$–2$\frac{1}{4}$ in) long, each individual flower about 3 mm. Fruit approximately 4 mm long, two-celled and two-seeded. Easily distinguished from buck's-horn plantain (*P. coronopus*) by its leaves and fruit. Grows in saltmarshes, and along estuary margins, also cliff ledges, rocky coasts, and in meadows close to the sea. In flower June to August and widely distributed around the coasts of the British Isles. The caterpillar of the glanville fritillary butterfly (*Melitaea cinxia*) (which occurs only on the Isle of Wight) shows a preference for the sea plantain as its food plant. The caterpillars of the ground lackey moth (*Malacosoma costrensis*) (best known on salt marshes adjacent to the Thames and the Medway) also feed on sea plantain and are fond of sunning themselves on the plant.

Sea Wormwood (*Artemisia maritima*)
Family Compositae (120) **Fig. 16(1)**

A locally common and pungently aromatic peren-nial herb with an erect, ascending or prostrate habit. Its flowering-shoots are branched above, 20–50 cm (7$\frac{3}{4}$–19$\frac{1}{2}$ in) in height, and usually covered with a silvery-white down. The leaves are 2–5 cm ($\frac{3}{4}$–2 in) long, mostly twice-pinnate with narrow linear leaf-lets of about 1 mm wide, woolly on both sides, and borne on each side of the stem. The flowerheads are small (1–2 mm in diameter), few, and crowded into short, drooping, panicle-like spikes; florets are reddish, or yellowish-orange. Flowering period is August to September. The caterpillars of several moths feed on sea wormwood between August and June, including the Portland moth (*Acetebia pra-ecox*), the coast dart (*Euxoa cursoria*), and the Essex emerald (*Euchloris smaragdaria*) (which seems con-fined to saltmarshes in Essex). The ground lackey (*Malacosoma castrensis*) (which is perhaps best known on the estuarine saltmarshes of the Thames and the Medway) lays her eggs around the stems of sea wormwood; the caterpillars later feed, and sun themselves, on the plant. It grows, and is locally common, on the drier parts of saltmarshes and sea walls around most of the British Isles, but seems not to have been recorded in the extreme north. In Ireland very local on east and west coasts.

Sea Poa or Sea Manna Grass (*Puccinellia maritima*)
Family Gramineae (146) **Fig. 16(6)**

This somewhat tufted, stoloniferous perennial is 10–80 cm (4–31$\frac{1}{2}$ in) in height. Its leaves are smooth and narrow, and often inrolled at margin. Panicle up to 25 cm (10 in) and narrow with two to three very slender and more or less erect branches ema-nating from each node. The spikelets are 7–12 mm long and usually comprised of six to ten florets. It

Fig. 15   1. Sea wormwood (*Artemisia maritima*). 2. Sea arrow-grass
(*Triglochin maritima*). 3. Sea plantain (*Plantago maritima*). 4. Sea purslane (*Halimione portulacoides*). 5. Sea clover (*Trifolium squamosum*).
6. Sea poa (*Puccinellia maritima*).

flowers during June and July and is widely distributed around the British Isles where it is common, and at times dominant in saltmarshes and muddy estuaries. The reflexed poa (*P. distans*) which is not stoloniferous, and whose spikelets are usually between 4 and 6 mm and have only three to six florets, are borne on lanceolate to triangular panicles of up to 15 cm (6 in).

Sea Arrow-grass (*Triglochin maritima*)
Family Juncaginaceae (125) **Fig. 16(2)**

Its common name 'arrow-grass' is derived from the arrow-like shape of the ripe fruiting-body of another member of the genus, the marsh arrow-grass (*T. palustris*), which grows in freshwater marshes inland. Sea arrow-grass is a rather stout herbaceous perennial whose leafless flowering-stems are up to 30 cm (12 in); it has a tufted habit. All the leaves are radical, linear, not furrowed, up to 3 mm wide, and more or less semi-cylindrical in section. The small inconspicuous yellowish-green flowers are in a long, loose raceme, and borne on pedicels of 1–2 mm. Flowering period is July to September. Habitat is saltmarsh turf, and other grassy places on rocky coast-lines. It is distributed around the entire coast of the British Isles wherever conditions are suitable.

Sea Clover (*Trifolium squamosum*)
Family Papilionaceae (51) **Fig. 16(5)**

The habit of this somewhat pubescent annual is prostrate, then ascending and usually between 15 and 45 cm (6–18 in) tall; stems do not root at nodes. The trefoil leaves have narrowly-obovate leaflets of 10–25 mm, and pedicels of up to 10 cm. All flower-heads are terminal, oval-shaped, and tightly crowded with numerous tiny pale pink flowers. Calyx-tube bell-shaped, glabrous, and of a leathery texture. In flower June to July. Occurs but locally in salt-marshes, coastal turf, and meadows adjoining tidal estuaries. Distribution mainly along the south coast, then north to Carmarthen and Lincolnshire, also found in Caernarvon and Lancashire.

Sea Purslane (*halimione portulacoides*)
Family Chenopodiaceae (34) **Fig. 16(4)**

The mealy coating of this very small shrub (usually up to 80 cm, $31\frac{1}{2}$ in, tall) attracts one's immediate attention. Its stems are decumbent with ascending branches of about 45 cm (18 in) in height, and others which are straggly or interweaving, and possibly longer. The lower elliptic leaves are opposite and on short petioles of 5–10 mm; upper leaves are spoon-shaped or oval, thickish, mainly opposite, and taper narrowly to stem. The dense clusters of tiny yellowish-green flowers are in compound terminal spikes or in axillary clusters. It is in flower from July to September, and grows in saltmarshes, along the edges of creeks and pools, usually flooded at high tide. Common and locally abundant in England north to Cumbria and Northumbria; also occurs in parts of Scotland and Ireland.

Tassel Pondweed (*Ruppia maritima*)
Family Ruppiaceae (129) **Fig. 17(6)**

This slender perennial is one of the few flowering plants which can withstand being submerged in salt water. Its stems are up to 30 cm (12 in) long and many-branched. The thread-like leaves are up to 0.5 mm wide, light green, and have a narrow basal sheath. Its greenish flowers, which rise above the water, grow at the end of long stalks which emanate from base of leaves. A locally common plant which grows in saltmarsh pools and brackish ditches. It flowers from July to September and is distributed around most of our coastline, but decreases in frequency northwards, occurs up to Shetland.

Coiled pondweed (*Ruppia spiralis*) is a much rarer species than *R. maritima* but of similar general appearance. The leaves are dark green and wider, up to 1 mm. Often the flower-stalks spiral after flowering, thus submerging the fruit. It also grows in brackish coastal ditches and on saltmarshes; the flowering period is July to September. It has a fragmented distribution around the coasts of the British Isles up to Orkney and Shetland.

Eel Grass (*Zostera marina*)
Family Zosteraceae (127) **Fig. 17(7)**

Once a common perennial on mud flats around our coasts, but during the early 1930s it was attacked by a disease which during just a few years destroyed large expanses of the plant, and consequently deprived many species of wintering wildfowl of their staple food; especially the brent goose (*Branta bernicla*) which feeds on it almost exclusively. Fortu-

Fig. 16 1. Shore orache (*Atriplex littoralis*). 2. Sea rush (*Juncus maritimus*). 3. Sea club-rush (*Scirpus maritimus*). 4. Common cord-grass (*Spartina × townsendii*). 4a. Close-up of *S. × townsendii*. 5. Cord-grass (*Spartina maritima*). 5a. Close-up of *S. maritima*. 6. Tassel pondweed (*Ruppia maritima*). 7. Eel grass (*Zostera marina*). 8. Eel grass (*Zostera angustifolia*).

nately eel grass is now showing signs of recovery. The plant has rhizomes of 2–5 mm thick, and produces flat, dark green, rather grass-like leaves which on sterile shoots are usually up to 50 cm (20 in) long and 5–10 mm wide; those produced on fertile shoots are shorter and relatively narrower. The flowering-stems are much branched and up to 60 cm (24 in) long, they bear spikes (9–12 cm, $3\frac{1}{2}$–$4\frac{1}{2}$ in) long of petal-less flowers which are enclosed in a leaf-like sheath. Flowering period June to September. It is one of the few British flowering plants that can endure being regularly submerged by the sea. It is rarely found in estuaries, and prefers the gravelly and muddy regions on the middle and lower shores. It occurs around our coasts, becoming more uncommon northwards. Sometimes the brown seaweed *Scytosiphon lomentaria* attaches itself to eel grass.

The narrow-leaved eel grass (*Z. angustifolia*) (Fig. 17, 8) has leaves on sterile shoots of up to 30 cm (12 in) long and only 2 mm broad; leaves of fertile shoots 4–15 cm ($1\frac{1}{2}$–6 in) long and 2–3 mm wide. Flowering stems are pale green to white, 10–30 cm (4–12 in) long and branched. It grows on estuarine mudflats and in shallow pools, from middle to lower shore. It has a scattered distribution around our coast north to Orkney.

Dwarf eel grass (*Z. nolti*) is, as its name suggests, very small, up to 15 cm (6 in). The thread-like leaves are one-veined, up to 12 cm ($4\frac{1}{2}$ in) long and up to 1 mm wide. Flowering-stems usually branched; flowering period June to October. It grows on mudbanks in estuaries and creeks from middle to lower shore. It is locally common around our coasts where conditions are suitable, and occurs as far north as Inverness; much less common along west coast.

Common Cord-grass (*Spartina × townsendii*)
Family Gramineae (146) **Fig. 17(4 and 4a)**

A stout, erect perennial whose flower-stem grows to between 50 and 130 cm (20–51 in). It often forms large patches of small tufts each up to about 40 cm (16 in) high. The leaves are yellowish-green to pale green, very stiff, spreading, and more or less horizontal, about 8 mm wide, and taper gradually from base to a long slender point; they barely reach up to the inflorescence which is between 10 and 25 cm (4–10 in) and are usually comprised of four to five spikes. Each spike contains several stiff, over-

lapping spikelets; these are yellowish-brown, 15–20 mm long and tightly pressed. Flowering period is June to August. It is a very abundant species on tidal mudflats, and has been planted to help bind the mud in some areas. It occurs along the south coast, parts of the west coast, and elsewhere.

Cord-grass (*S. maritima*) (Fig. 17, 5 and 5a) is far less stout than *S. × townsendii* with shorter flower-stems of 10–50 cm (4–20 in). The leaves are softer, often purplish, about 4 mm wide, and taper to a somewhat stout point. Inflorescence is on average much shorter, between 6 and 12 cm ($2\frac{1}{2}$–5 in) and with two to three erect spikes. Spikelets 10–14 mm long, golden or yellowy-brown. In flower August to September. A native on tidal mudflats, occurring along south coasts, from south Devon eastwards, and north to Lincolnshire; also Glamorgan and probably elsewhere in south and south-west Wales.

Sea Club-rush (*Scirpus maritimus*)
Family Cyperaceae (145) **Fig. 17(3)**

This stout glabrous perennial is one of our commonest maritime species. It grows from 30 to 100 cm (12–40 in) and is propagated by suckers. The stem is three-angled, becoming rough towards tip; they grow in tufts and bear leaves at the base, which are up to 10 mm wide, keeled and have rough margins. The flowers are borne in a dense, terminal cluster containing two to five spikelets; each spikelet is 10–20 mm long, ovoid, and reddish-brown. Bracts are leaf-like and bristly, the larger ones longer than the inflorescence. Flowering period is July to August. Locally abundant in shallow water along the muddy edges of tidal rivers, also in ponds, and along ditches near the sea. It is only rarely found inland. It occurs around our coasts as far north as Easter Ross and Isle of Lewis.

Sea Rush (*Juncus maritimus*)
Family Juncaceae (136) **Fig. 17(2)**

This often grows in a continuous fringe along salt-marshes where it can be dominant over large areas above the high-water mark of spring tides. It is an erect, very tough and rather tufted perennial whose deep roots form a densely-matted condition and therefore have a stabilising effect on the substrate.

The light green stems are 30–100 cm (12–20 in) tall, and smooth in new growth; the pith is continuous. The leaves are acutely pointed and emanate from near stem base only; the lowest ones have shiny brown sheaths. Inflorescence is a loose terminal cyme of pale petal-less flowers with ascending stems which are shorter in length than the sharply-pointed bracts. Flowering period July to August. Occurs on saltmarshes and is abundant on coasts up to Inverness-shire.

Shore Orache (*Atriplex littoralis*)
Family Chenopodiaceae (34) **Fig. 17(1)**
Often referred to as the narrow-leaved sea orache. This deep-rooted and rather mealy annual has stoutish, erect, and much-branched stems; it grows up to 100 cm (20 in) tall. Leaves narrow, linear to linear-oblong, and usually entire but sometimes dentate. Lower leaves have short stalks, upper ones are sessile and very narrow. The spike-like inflorescence is up to 20 cm (8 in) with basal leaves only, it bears clusters of small petal-less greenish flowers during July and August. Usually grows on muddy saltmarshes, or brackish marshes near the sea. Occurs all around the coast, but somewhat local in Scotland and Ireland.

Glasswort or marsh samphire (*Salicornia europaea*).

Glasswort (*Salicornia europaea*)
Family Chenopodiaceae (34)
Also referred to as marsh samphire. An erect herbaceous annual of unusual structure which usually grows between 15 and 30 cm (6–12 in) tall. The stems are bright green and much-branched; they appear to be leafless, but the succulent leaves are set in opposite pairs and joined together along the edges to form segments which completely envelop the stem. They later turn yellowish-green and are finally flushed reddish-pink. Terminal spikes usually 15–50 mm, slightly tapering, and comprised of five to nine fertile segments. Tiny, green, petal-less flowers appear in threes either at, or near, the junctions of the jointed branches. It is in flower during August. It grows locally on open sandy mud and in saltmarshes. It occurs along the south and west coasts of England. There are many related species, all of similar appearance and not always easy to differentiate.

Creeping or perennial glasswort (*S. perennis*) which is a rather woody perennial produces tufts of ascending or decumbent stems up to 30 cm (12 in).

The terminal spikes are 10–20 mm, and on average are smaller than those of *S. europaea*. It flowers August and September and grows on saltmarshes and gravelly shores, principally in southern England.

Twiggy glasswort (*S. ramosissa*) is an erect or sometimes prostrate annual growing up to 40 cm (16 in). It has a bushy appearance and is much-branched; the tapering terminal spikes are 10–30 mm with four to nine fertile segments. It ultimately turns deep purplish-red, giving the plant an almost bead-like composition. It is common and widespread in saltmarshes on firm bare mud. It occurs in suitable areas along the eastern and southern coasts of England, and south Wales.

Bushy glasswort (*S. dolichostachya*) is an erect, many-branched annual 10–40 cm (4–16 in) tall with a bushy habit. Initially dark green, then paling to sullied yellow and finally brownish. Terminal spikes are usually 50–120 cm (20–47 in) and as a rule decidedly tapering, with 12–25 fertile segments. Flowering period is July to August; it grows on saltmarshes in firm muddy areas or in muddy sand. Occurs along our coasts from Lancashire south to Devon, and from Kent north to Easter Ross; also Ireland in west Galway and along the south and east coasts. There are other related species which occur locally and always in similar habitats.

Knotgrass (*Polygonum aviculare*)
Family Polygonaceae (79)

A very common annual with an erect or partially prostrate, straggling habit. It is abundant around farmland and waste places, but also occurs in salt marshes and other coastal habitats. The slender wiry stems are much-branched and up to 30 cm (12 in) in length, often much longer. The stem-leaves are lanceolate to ovate-lanceolate, up to 5 cm (2 in) long and 1.5 cm ($\frac{1}{2}$ in) wide; those of main stem are at least twice as long as the ones on flowering-stems. Very small, solitary, usually pink or white petal-less flowers are borne in leaf axils on the stem; they are barely 2.5 mm and give the plant its 'knotted' appearance. The flowering period extends from April to October. It is a food plant for the caterpillars of numerous moths, including the red sword-grass (*Xylena vetusta*) from May to July; the red twin-spot carpet (*Xanthorhoë spadicearia*) June and July, and also September and October; the fan-foot (*Zanclognatha tarsipennalis*) and the shears (*Hada nana*) July and August; the mottled rustic (*Caradrina morpheus*) August and through the autumn; and the satin wave (*Sterrha subsericeata*).

Other related species include the very rare perennial sea knotgrass (*Polygonum maritimum*) whose much thicker, larger grey leaves have rolled-down margins. It occurs locally in the Channel Isles but may well be extinct in Britain. The slender knotgrass (*P. raii*) is a straggling and prostrate annual whose shiny reddish-brown stems have a woody base, and which occurs on fine shingle, and sandy shores. This is now very uncommon but grows in suitable habitats along the south and west coasts up to the Hebrides.

Marsh Mallow (*Althaea officinalis*)
Family Malvaceae (37)

This has an erect habit and a stout woody stem of 60–120 cm (24–47 in) which is densely pubescent, almost velvety; often entire but sometimes slightly branched. The large leaves are heart-shaped to egg-shaped or almost round, but with three to five shallow lobes and irregularly-toothed margin; upper leaves more deeply and acutely lobed than lower ones. All leaves are velvety on both sides and have fan-like folds; lower leaves large, 3–8 cm ($1\frac{1}{4}$–3 in) across. The flowers are pale rosy-pink and 2.5 to 4 cm (1–$1\frac{1}{2}$ in) across; they appear in upper leaf axils and are usually borne in small clusters of about three, but sometimes occur singly; their stalks are shorter than the leaves. The entire plant has a soft velvety feeling when handled, and its flower clusters bear a certain similarity to those of the hollyhock. Period of flowering is August to September. Marsh mallow is a locally common perennial and native along the margins of both salt and brackish marshes, and other suitable situations close to the sea. It occurs around our coasts up to Northumberland in the east, Dumfries and Kirkcudbright in the west, also Isle of Arran and Jersey.

Dwarf Mallow (*Malva neglecta*)
Family Malvaceae (37)

An herbaceous annual with densely hairy, decumbent stems of 15–60 cm (6–24 in), but often the central stem is erect. Leaves long-stalked, 4–7 cm ($1\frac{1}{2}$–$2\frac{3}{4}$ in) in diameter, somewhat pubescent, heart-shaped with five to seven shallow lobes, and margins acutely scalloped. Flowers whitish and veined with some shade of lilac, 1.8–2.5 cm ($\frac{3}{4}$–1 in) across, and arranged in irregular racemes. In flower June to September. Grows along the drift line but not confined to coastal regions, also found inland along roadsides and in waste places. Occurs throughout the British Isles with the exception of Outer Hebrides and Shetland; is most frequent along southern coasts.

Small Mallow (*Malva neglecta*)
Family Malvaceae (37)

This is a small version of *M. neglecta* but the flowers do not grow in dense clusters, and are only 0.5 cm in diameter. Petals range in colour from pale rose-pink to whitish. In flower June to September. It is an introduced annual which occurs along many foreshores and also on waste places inland. It is of local distribution in England and Wales, also Scotland north to Aberdeen.

Sea Aster (*Aster tripolium*)
Family Compositae (120)

A locally common herbaceous perennial with stout, erect stems usually about 30 cm (12 in) high but may reach up to three times that dimension. Its smooth succulent leaves are dark green; basal ones broadly lanceolate, narrowing into a long stem; stem-leaves are linear or broadly linear, tapering abruptly at each end. Flowerheads are 8–20 mm in diameter and occur in loose clusters; the ray florets are some shade of bluish-purple (rarely white), the disk florets are tubular and yellow; very much like the cultivated Michaelmas daisy. In flower July to October and common in saltmarshes, along the edges of mudflats and creeks, also occurring on sea-cliffs. Widely distributed around our coasts, but in northern England and Scotland it is most likely to be found in the vicinity of estuaries.

Sea Beet (*Beta vulgaris* ssp. *maritima*)
Family Chenopodiaceae (34)

Sometimes referred to as wild spinach because the basal leaves are regarded by some as even tastier than garden spinach. This very common herbaceous perennial grows up to 90 cm (35 in), the lower part of its stem is often prostrate then ascending with that tall and familiar succulent growth; it is numerously branched and angular, giving the plant a somewhat floppy, pyramidal outline. The leaves are usually up to 10 cm (4 in) long, fleshy, often shiny, deep green, wavy, and broadly triangular to egg-shaped before narrowing into a leaf-stalk. From June to October it produces long leafy spikes bearing clusters of two to three small greenish flowers at regular intervals. Its habitat includes saltmarshes, and the edges of mudflats, along shingle beaches and rocky muddy shores, also on sea-cliffs. Occurs throughout coastal regions of the British Isles, less common in northern Scotland.

Sea Lavender (*Limonium vulgare*)
Family Plumbaginaceae (95)

This deeply tap-rooted herbaceous perennial is often referred to as statice, a former generic name.

Its flowering-stems are rather angular, erect, and usually up to 30 cm (12 in) high with corymbose branches emanating from above midway; the lowest of which are often barren. The strongly pinnate-veined leaves are 4 to 12 cm ($1\frac{1}{2}$–$4\frac{1}{2}$ in) long or more, and vary greatly in shape between broadly elliptic and broadly lanceolate, usually terminating in a short narrow point, and gradually narrowing into a long slender stalk; they form a basal rosette. The numerous flowering-stems are really scapes because all the leaves are radical; the small blue-purple flowers are borne terminally in spikes on a many-branched flattish-topped flowerhead, each spikelet bears two or three flowers, and these are set in two rows on the upperside of the spikes. The corolla is only 8 mm in diameter, and the blue-purple petals are broadish and rounded. Flowering period is July to October. This native perennial is often abundant or even dominant on muddy saltmarshes around the coasts up to Dumfries and Fife.

Lax-flowered sea lavender (*Limonium humile*) resembles *L. vulgare* but the flowering-stems usually begin to branch from below middle of main stem, and inflorescence is not corymbose; its long, lax spikes tend to be incurved. The leaves are oblong-lanceolate and obscurely pinnate-veined. In flower July and August. This native perennial occurs in muddy saltmarshes, and is distributed coastally up to Dumfries and Northumberland; can be found all around the coast of Ireland but becomes rare in the north.

Matted sea lavender (*Limonium bellidifolium*) has much smaller leaves than either *L. vulgare* or *L. humile*, only 1.5–4 cm ($\frac{1}{2}$–$1\frac{1}{2}$ in) and very few in each basal rosette; they die before end of flowering period. Flowering stems have several many-forked, barren, zig-zag branches near to base. Fertile spikes are only borne on uppermost branches, they are numerous and dense; spikelets are closely set in two rows on upperside of spikes. Corolla only 5 mm in diameter and pale lilac. Flowering period is July and August. Found on dried areas of sandy saltmarshes. Distribution seems restricted now to coasts of Norfolk and Suffolk.

Sea Milkwort (*Glaux maritima*)
Family Primulaceae (96)

An alternative name for this little semi-prostrate perennial is black saltwort. The stems are suberect

and usually about 15 cm (6 in) in height, but up to twice that dimension on occasions. Its smooth, fleshy, bluish-green leaves are darker on the upper surface, only 4–12 mm long, oval or elliptically oblong, and almost stalkless. The pale pink, petalless flowers are only 5 mm in diameter and borne singly in leaf axils; calyx is bell-shaped and flowerlike. Flowering period is June to August. Sea milkwort grows in patches on grassy saltmarshes and other suitably saline and sandy substrates along coasts and estuaries, even in splash-zone; also in rock crevices. A locally common species around all the coastline.

Wild chamomile (*Chamaemelum nobile*), ragwort (*Senecio jacobaea*), sea beet (*Beta vulgaris* ssp. *maritima*) and marsh mallow (*Althaea officinalis*).

Wild Celery (*Apium graveolens*)
Family Umbelliferae (75)

A strongly odorous, erect biennial with a stout, grooved, and leafy stem which grows up to 60 cm (24 in) in height. The radical leaves are pinnate with serrated lobes on each side of a common stem.

Segments range in size from 0.5–3 cm ($\frac{1}{4}$–1$\frac{1}{4}$ in), are triangular or diamond-shaped. Upper-stem leaves are smaller, stalkless, and divided into three leaflets. Its tiny, only 1 mm, stalkless, greenish-white flowers are borne in small terminal or axillary umbels. The plant smells very similar to cultivated celery and this should aid identification, but is *poisonous* and must on no account be eaten. Flowering period June to August. It has a general distribution, principally in coastal counties where it grows in a variety of damp places, alongside rivers and ditches, and on saltmarshes. Occurs locally becoming rarer northwards, and absent in extreme north of Scotland. Can be found in many coastal areas of Ireland.

Dittander (*Lepidium latifolium*)
Family Cruciferae (21)

Also called broad-leaved pepperwort. An herbaceous perennial with a long thick rootstock which produces many subterranean creeping stems. The habit is erect with stems of 50–130 cm (20–51 in) which are hairless and upperparts much-branched. Its long-stalked radical leaves are oblong with a coarsely-toothed margin, or pinnately lobed; leaves of lower stem are similar but with shorter stalks; upper-stem leaves are stalkless but do not clasp stem, 5–10 cm (2–4 in) long by 1–2 cm ($\frac{1}{2}$–$\frac{3}{4}$ in) wide and ovate to ovate-lanceolate. Topmost leaves have white margins at apex and are very bract-like. The small creamy-white flowers are 2.5 mm in diameter and borne in dense cymes, the large inflorescence being conical in outline. Petals may be twice as long as the white-margined sepals. The fruit is only half the length of fruit-stalk. Flowering period June to July. Grows in saltmarshes and damp sand. Occurs from Norfolk and Wales southwards; also coast of north-east England and southern Ireland.

Narrow-leaved pepperwort (*Lepidium ruderale*). This somewhat offensive-smelling herb may be annual or biennial, its slender tap-root produces a single, more or less erect, stem of 10–30 cm (4–12 in) which may or may not be slightly hairy. Its branches are upcurving and emanate from upperpart of stem. The long-stalked basal leaves are 5–7 cm (2–2$\frac{3}{4}$ in) long and deeply pinnate; the narrow segments are themselves often further pinnately-divided or lobed. Leaves of lower-stem are pinnately-divided into narrow segments; those higher up stem are stalkless, margins entire, up to 20 × 2 mm and blunt. The tiny inconspicuous greenish flowers are usually petal-less; the fruit not more than 2 mm wide. Flowering period May to July. Usually grows in waste places and along roadsides near the sea. It is widely distributed throughout England but most frequent in East Anglia. Rare in Scotland and seemingly not recorded in Channel Isles and Ireland.

Greater Sand-spurrey (*Spergularia media*)
Family Caryophyllaceae (30)

Also known as saltmarsh sand-spurrey. This perennial has a prostrate habit with many stout flattish shoots of up to 30 cm (12 in), sometimes longer. The narrow fleshy leaves have a flat upper-surface but are rounded beneath, they are 1–2.5 cm ($\frac{1}{2}$–1 in) long by 1–2 mm wide, more or less acute with a horny tip, and usually void of hairs. Flowers small, only 7.5–12 mm in diameter and borne in a lax cyme in which the stems are opposite and more or less of equal length. The oval petals are pink with a white base and 4.5–5.5 mm long; the sepals are a little shorter. Flowering period is June to September. Grows in muddy sandy saltmarshes and similar coastal habitats. Greater sand-spurrey occurs generally around our coasts where the conditions are suitable.

Lesser sand-spurrey (*Spergularia marina*). As its name suggests, this herbaceous annual is smaller than *S. media*, having stems up to 20 cm (8 in). It has the same prostrate habit with branched stems; the narrow, fleshy, yellow-green leaves are set opposite. The flowers too are smaller, 6–8 mm in diameter and borne in a cyme whose stems are opposite but not necessarily of equal length. The five small, bluntly oval, pink petals are usually white at the base, only 2.5–3 mm long, and shorter than the sepals. Lesser sand-spurrey prefers the drier areas on muddy sandy saltmarshes and similar coastal habitats. It is in flower June to August, and occurs in suitable localities around our coasts; occasionally found on saline areas inland; Cheshire for example.

Sea Hard-grass (*Parapholis strigosa*)
Family Gramineae (146)

A slender glabrous annual of 15–40 cm (6–15 in). The stems are freely-branched, usually decumbent at base, and bent above; very seldom erect. Its leaves are short, roughish, and somewhat leatherish with inrolled margins; uppermost sheaths are inflated, and away from base of spike at maturity. The green spikelets are 4–6 mm long and rounded on back. Flowering period June to August. It is widely distributed, but local, around our coasts up to W. Lothian and Mull. Grows in grassy saltmarshes and waste places by the sea.

Lesser sea hard-grass (*P. incurva*) as its name suggests is smaller than *P. strigosa*, usually only 5–8 cm (2–3 in) tall, and grows amongst other coastal vegetation. Unlike *P. strigosa* the uppermost sheaths are more or less inflated, and reach base of spike or enclose base at maturity. In many other respects lesser sea hard-grass is similar to *P. strigosa*. Flowering period June to July. Often found in bare places and more usually on muddy shingle. Occurs in southern parts of England from Bristol Channel to Dorset, and Kent to Norfolk.

Herbaceous Seablite (*Suaeda maritima*)
Family Chenopodiaceae (34)

This common annual is also referred to as common or annual seablite. Its habit may be erect and up to 30 cm (12 in), semi-erect, or even prostrate and trailing. The branched stems are glaucous and often have a reddish tinge in late autumn; they bear numerous succulent, linear, pointed leaves which taper below and are somewhat semi-cylindrical in shape; usually 3–25 mm long by 1–2 mm wide. Its tiny green flowers occur singly or up to three together in small axillary cymes; they are far from conspicuous. Flowering period is July to October. Grows on seashores and in saltmarshes around most of our coastline, but seemingly not recorded in Berwick, Banff, Caithness and west Sutherland. It is a food plant for the caterpillar of the sand dart moth (*Agrotis ripae*).

Shrubby Seablite (*Suaeda fruiticosa*), although a related species to *S. maritima*, is a small evergreen shrub of 40–120 cm (16–47 in) with suberect or ascending branches which may be up to 5 cm (2 in) in diameter, and is much-branched. The leaves are evergreen, thick, linear, 5–18 mm by 1 mm, rounded at tip and alternate. Flowering period and flowers are as for *S. maritima*. This uncommon perennial occurs on saltmarshes and stable shingle shores principally along the south coast, and the east coast south from Lincolnshire.

Mud Rush (*Juncus gerardii*)
Family Juncaceae (136)

This medium-sized tufted perennial has the typical rush habit. Its stiffly-erect flowering-stems are up to 30 cm (12 in). The dark green leaves from 10 to 20 cm (4–8 in) are long and channelled. The flowers are borne in a compound and somewhat cymose terminal panicle, much longer than the bracts. Flowering period June to July. Usually grows in saltmarshes where it is often abundant or even locally dominant. It occurs around our coasts up to Shetland; also Ireland.

Initial figures refer to locations on Map 3.

1. THE SEVERN ESTUARY (GWENT/GLOUCESTERSHIRE/AVON)

The Severn is one of the longest estuaries in Britain and is best treated in two parts – the 'Welsh' Severn bordering the counties of Gwent and Mid-Glamorgan; and the stretch along the English coastline. One of the most important habitats associated with the estuary is the low-lying pasture. Such a habitat exists at Slimbridge, where the Wildfowl Trust has a reserve, and at Aylburton and the Somerset Levels near Bridgwater.

Waders are spread over most of the estuary, but wildfowl tend to concentrate around the Slimbridge area, where Bewick's swans and white-fronted geese are common during winter. In fact the estuary supports nationally important numbers of both these species as well as black-tailed godwit, shelduck, wigeon, teal, shoveler, ringed plover, grey plover, lapwing, knot, dunlin, whimbrel, curlew, redshank and turnstone.

Major wildfowl roosts occur at Slimbridge and Burnham whereas waders concentrate on the Caldicot Levels and Bridgwater Bay area. Breeding wildfowl or waders are few, but the islands of Flatholm and Steepholm have significant colonies of nesting gulls.

2. TAW/TORRIDGE (NORTH DEVON)

The Taw Estuary from Barnstable and the Torridge from Bideford join and enter the sea at Appledore. The National Nature Reserve of Braunton Burrows lies on the northern boundary of the outer estuary.

The Taw is by far the larger of the two and extensive areas of saltmarsh exist at intervals along the southern shore, best viewed from the railway which passes close by. The Torridge is narrower and there is little saltmarsh. It is also subject to major recreational disturbance, as is the Taw in fact. This is likely to increase with proposals for a new marina at Appledore.

Five species occur in nationally important numbers. These are curlew, golden plover, oystercatcher, ringed plover and sanderling.

3. CAMEL (CORNWALL)

The Camel is subject to major disturbance from recreational activities throughout most of the year, in particular during the summer when scores of visitors flood into the popular resorts of Padstow to the west and Rock on the opposite side of the estuary. Sailing and water-skiing have always been popular in the area and lately wind-surfing.

White-fronted geese graze the marshes near the quaint village of Chapel Amble, whereas wading birds like turnstone, oystercatcher and golden plover tend to use the outer estuary.

There are many pleasant walks in the area, especially the coastal one from Padstow northwards towards Stepper Point, along which tamarisk trees flourish.

4. FAL COMPLEX (CORNWALL)

Carrick Road extends northwards from Falmouth Bay and is a focus for four separate estuaries; these

Map 3   Estuarine habitats.

White-fronted geese (*Anser albifrons*) on the New Grounds at Slimbridge.

are the Fal, Truro, Tresillian and Restronguet Creek. Few saltmarshes exist on any of them and each is essentially muddy in character.

The nationally important species wintering on the complex are curlew, spotted redshank, greenshank and black-tailed godwit. Other species present in numbers include shelduck, wigeon, dunlin, redshank and lapwing.

There are many threats to the wildlife of the area, including increased pollution from local industries and shipping.

5. TAMAR COMPLEX (CORNWALL/ DEVON)

The Tamar complex consists of five separate estuaries, some of which are in Cornwall and the others in Devon, although it must be said that the Tamar itself is partly in both counties.

The only major area of saltmarsh is along the inner reaches of the Lynher. Generally, the intertidal flats are muddy in nature. Like several of the larger estuaries along the south coast, disturbance from recreational activities is a major influence on the distribution and number of birds feeding or roosting on the complex.

Six species are present in nationally important numbers; most of the British wintering population of avocets feed on the Tamar and Tavy. Black-tailed godwits exist throughout the complex.

6. EXE ESTUARY (SOUTH DEVON)

Birds feed on most of this estuary, which is the largest in the south-west. The main roost for wildfowl and waders is on Dawlish Warren; 200 hectares (500 acres) of it comprise a local nature reserve. As many as 20,000 birds can be seen at high tide from a hide erected on the site. These include Brent geese,

black-tailed godwits, dunlin and wigeon. The area is subject to intense pressure from holidaymakers in the summer, however; watersports are particularly popular.

Tamarisk (*Tamarix anglica*) growing alongside the Camel Estuary.

7. ARNE (DORSET) RSPB RESERVE

The Arne reserve, which comprises nearly 520 hectares (1,300 acres) is approached from the A351 Wareham to Swanage road, turning left at Stoborough village. The Reception Centre and car parks are located 3 km (2 miles) along the Arne road.

The site consists of large areas of heath covered with heather and gorse, together with expanses of deciduous and coniferous woodlands, reedbeds and saltmarsh. The heath provides a nesting ground for Dartford warbler, stonechat and nightjar; the woodland for green and great-spotted woodpecker and blackcap; the reedbeds for reed bunting and reed warbler; and the saltmarsh for redshank.

Winter visitors include goldeneye, red-breasted merganser, wigeon, black-tailed godwit and spotted redshank. Over 400 species of plant and 800 species of moths and butterflies have been recorded at the reserve.

Dawlish Warren, a local nature reserve beside the Exe Estuary, and an important wader roost at high tide.

Between April and August, part of the reserve is only open to visitors on certain days in the week, and then by conducted tour booked in advance from the warden at Syldata, Arne, Wareham BH20 5BJ. The Shipstal part of the reserve is open all year and does not require a permit. The Shipstal Point nature trail opens from late May to early September; it starts opposite Arne Church at the beginning of the Shipstal bridleway. There is a charge for non-members.

8. KEYHAVEN AND PENNINGTON MARSHES (HAMPSHIRE)

The reserve is mainly composed of saltmarsh and mudflat, although it does include part of the Hurst Castle shingle spit. The site is well known for its breeding terns, including the little, common and sandwich species. A large colony of black-headed gulls also nests on the site.

The area is heavily disturbed during the summer –

wind-surfers in particular like it, mainly because the waters behind the spit are fairly sheltered, plus the fact that they can 'drive' their gear to within 50 metres of the surfing area. The approach to the spit can get very cluttered with vehicles during the holiday season and visitors might prefer to park in Keyhaven village and walk to the site.

It is also a popular area for sailing, but boats must not visit the nesting area during the breeding season. Perhaps the best view of the reserve is from the public footpath along the sea wall.

9. SOUTHAMPTON WATER (HAMPSHIRE)

Although the banks of almost all the estuary are highly developed, birds tend to concentrate on particular sites such as an area of marsh at Fawley and near the mouth of the river Hamble. On the inner part of the estuary, waders roost on Eling Marsh to the north-west and Dibden Bay near Hythe, where wildfowl also congregate. Huge numbers of gulls roost on the estuary during the winter.

10. LANGSTONE HARBOUR (HAMPSHIRE) RSPB RESERVE

The reserve covers nearly 560 hectares (1,400 acres)

The upper reaches of Chichester Harbour beside the A3023 near Langstone.

of mudflats and saltmarsh islands and is a most important site for wintering wildfowl. There are no visiting facilities, but good views can be gained from the coastal footpath between Langstone and Chichester Harbours. The site is located to the west of the A3023 between South Hayling and the A27 Chichester/Fareham road.

Some 6,000 dark-bellied Brent geese, teal, wigeon, shelduck, dunlin, curlew, redshank, black-tailed godwit, black-necked grebe and red-breasted merganser can be seen during winter. Dunlin are by far the most numerous wading birds, with upwards of 20,000 present on some occasions. The black-tailed godwits account for 10–15 per cent of the British wintering population.

There are no charges at the reserve and information regarding best viewing times and parking spaces can be obtained from the Warden, 6 Francis Road, Purbrook, Portsmouth P07 5HH.

11. FARLINGTON MARSHES (HAMPSHIRE)

Situated in the north-west corner of Langstone Harbour, and clearly visible from the main A27 road, this (300-acre) reserve contains a range of habitats, including saltmarsh, brackish pools, scrub and freshwater marsh.

Wildfowl and wader numbers are impressive throughout winter, with Brent geese and black-tailed godwit topping the importance list. Plant species found on the reserve include sea clover and golden samphire.

Access is by gravel track off the A27/A2030 roundabout (main sign 'Southsea') under the M27 flyover. Further details are available from the Hants and Isle of Wight Trust.

12. CHICHESTER HARBOUR (SUSSEX/HAMPSHIRE)

Unrivalled in both size and shape along the whole coast of southern England, Chichester Harbour consists of four major sections, with extensive areas of sand and mudflats in the outer estuary, compared with a predominantly saltmarsh inner area, especially either side of the Ministry of Defence owned Thorney Island.

The largest roosts are on Thorney and Hayling

islands and eight species occur in nationally important numbers, including dark-bellied Brent geese, shelduck, grey plover and black-tailed godwit.

Like most of the south-coast estuaries, disturbance from water-sport enthusiasts is acute at certain times of the year, but otherwise the harbour is relatively undeveloped.

13. PAGHAM HARBOUR (SUSSEX)

Although small compared to other local estuaries, Pagham is nevertheless a superb area for watching birds, especially on migration. Over 400 hectares (1,000 acres) are designated a local nature reserve.

There are extensive areas of saltmarsh on which typical maritime plants flourish, including glasswort and sea purslane. Birds roost mainly on shingle banks within the estuary, but high tides cause them to venture onto fields, especially on the western side of the reserve. A large variety of both waders and wildfowl have been recorded, although none occurs in internationally important numbers. Of national significance are grey plover, shelduck, black-tailed godwit and dark-bellied Brent goose. A few avocets are frequently seen together with over 50 ruff during most winters.

The site is an excellent area for birds and the path around it makes the whole reserve accessible. Since sailing and other water-sports are not allowed within the estuary, there is little disturbance.

14. SOUTH SWALE (KENT)

The Kent Trust for Nature Conservation manages this 200-hectare (500-acre) reserve covering some 5 km (3 miles) of sea wall and foreshore from Faversham Creek to Sportsman Inn, a mile north of the village of Graveney, where parking is available.

The reserve is predominantly tidal mudflats with large patches of eel grass attracting several species of wildfowl in winter, notably wigeon and dark-bellied Brent geese.

15. ELMLEY MARSHES (KENT) RSPB RESERVE

Over 1,320 hectares (3,300 acres) of grazing marshes, shallow pools bordered by saltmarsh, form an ideal habitat for wintering wildfowl and breeding waders.

The reserve is reached by following the A249 Sheerness road and taking the track turn-off 1.5 km (1 mile) beyond Kingsferry Bridge. Two miles of rough track lead to Kingshill Farm where cars can be parked about a mile from several hides which overlook flooded lagoons and mudflats. Limited parking close to the hides exists for disabled or elderly visitors.

Very large numbers of wigeon, teal, mallard, shelduck and whitefronted geese winter here and the marshes support considerable numbers of dunlin, black-tailed godwit, curlew and redshank. Breeding birds include shoveler, redshank and pochard.

The reserve opens throughout the year on certain days of the week between 10 am and 6 pm. Non-member visitors are charged a small fee and more detailed information is available from the warden, Kingshill Farm, Elmley, Sheerness, Isle of Sheppey ME12 2RN.

16. MEDWAY (KENT)

The extensive mudflats and saltmarsh islands of the Medway are of exceptional interest ornithologically, with no fewer than seven species occurring in internationally significant numbers, especially dabbling duck. Major roosts exist on all the major saltmarsh island complexes, especially alongside Chetney and Barksore Marshes.

There are several threats to the birdlife of the area. Recreational activities such as sailing, water-skiing and shooting all play their part.

17. INNER THAMES (KENT/GREATER LONDON/ESSEX)

The Inner Thames has been subjected to considerable development over the years, and prior to the early 1960s pollution from chemical and organic effluent was a major problem. There has been an immense improvement over the last 20 years or so, with a corresponding increase in the bird population of the estuary.

Wildfowl common on the estuary include mute swan, shelduck and pochard. Dunlin top the wader list, but at certain times of the year significant numbers of redshank and ruff make an appearance. Wildfowl tend to concentrate near Barking, Aveley Marshes and further out on Mucking Flats, near Stanford le Hope. Waders are more widespread, preferring Rainham and Dartford Marshes, and lagoons at Cliffe and near West Thurrock Power Station.

18. TWO TREE ISLAND (ESSEX)

Two Tree Island is part of a National Nature Reserve which the Essex Trust manages with the Nature Conservancy Council. Consisting of some 72 hectares (180 acres) in total, the reserve is situated on the eastern part of the island dominated by saltings and inter-tidal mudflats. These are feeding grounds for many wildfowl and waders, including Brent geese in winter.

The reserve is approached via the road bridge at Leigh-on-Sea station and there is ample car parking space available. A leaflet on the reserve is available from the Trust headquarters.

19. FOULNESS AND MAPLIN SANDS (ESSEX)

The extensive area covered by Foulness and Maplin Sands, together with Havengore, Rushley and Potten Islands, is one of the largest inter-tidal flats complex in Britain. Most of it is owned by the Ministry of Defence and as a result there is little

Cley Marshes Nature Reserve viewed from the Visitor Centre situated beside the main A149 road at Cley-next-the-Sea.

development. Ornithologically, the site is of immense importance, providing rich feeding and ample roosting sites for both wildfowl and waders.

No fewer than seven species occur in internationally important numbers, including dark-bellied Brent goose and bar-tailed godwit. The whole stretch of foreshore is used by the birds. There is also a significant breeding population of little, common and sandwich terns, ringed plover and black-headed gulls. Numerous other gulls can be seen in the area outside the breeding season, especially during autumn.

20. BLACKWATER (ESSEX)

The estuary offers a range of habitats to wildlife, ranging from mudflats to small islands and both salt and freshwater marsh. Wildfowl concentrate near Osea and Northey Islands with dark-bellied Brent geese and shelduck numbers being internationally significant. Waders tend to be found further out either side of Coldhanger Creek and near Tollesbury. Curlew and redshank numbers often exceed 2,000, with over a hundred pairs of the latter species breeding in the area.

21. NORTHEY ISLAND (ESSEX)

This 104-hectare (260-acre) reserve is managed under licence from the National Trust and is primarily a bird reserve consisting mainly of saltings. Breeding birds include redshank and shelduck. In winter, huge flocks of Brent geese feed on the site together with various ducks and waders. Black-tailed godwits are frequent visitors at passage times.

Special permission is needed to visit this reserve. This is obtained by writing to the Warden at Northey Cottage, Northey Island, Maldon, Essex.

22. FINGRINGHOE WICK (ESSEX)

Excavation work in the past has created a variety of habitats on this Essex Trust reserve, composed mainly of disused gravel workings and adjacent saltings on the Colne Estuary.

A shallow lagoon behind the sea wall attracts numerous birds, many of which nest nearby. Other interesting features include an extensive area of birch-oak wood, which has developed over the years on land previously dominated by willow scrub.

The reserve can be reached along a lane, signposted to South Green, off the road between Fingringhoe and Langenhoe. A car park is available behind the Information Centre, which also serves as the Trust's headquarters. Enquiries should be addressed to the Warden, Arch Hall, Fingringhoe, Colchester.

23. ORWELL AND STOUR (SUFFOLK/ ESSEX)

Both estuaries join and enter the sea at Harwich harbour and are best described as long and thin in structure with a balance of sandy and muddy sections. The major wildfowl and wader roosts occur on the upper reaches of the two; Freston on the Orwell and from Holbrook Bay westwards on the Stour.

The Orwell supports a variety of birds, but is perhaps more important for its wildfowl species – 18 in all, with several present in nationally important numbers. Like the Stour, there are several hundred non-breeding mute swans present during the summer.

The Stour, on the other hand, boasts internationally important numbers of no fewer than five species – redshank, shelduck, pintail, grey plover and black-tailed godwit.

24. BREYDON WATER (NORFOLK/ SUFFOLK)

Breydon Water is protected from the sea by the town of Great Yarmouth, the link between them being the narrow channel of the river Yare. It is established as a nature reserve and supports bean geese and Bewick's swans in nationally important numbers. Several rare species visit the area.

25. CLEY MARSHES NATURE RESERVE (NORFOLK)

The reserve, consisting of some 170 hectares (435

Black-tailed godwits (*Limosa limosa*) breed on Cley Marshes.

26. MORSTON AND STIFFKEY MARSHES (NORFOLK)

These marshes are of immense importance both for birds and flowers. Plants include sea aster and sea lavender which are in flower during late summer. Little and common terns breed on the shingle banks and during winter, Brent geese, wigeon and teal feed on the marshes.

Morston Marshes can be reached via the track to Morston Quay and by public footpaths along the landward edge of the marsh from Blakeney or Stiffkey car park.

There is access for vehicles to Stiffkey Marshes down a road called Greenway to the camp-site. The car park has room for over 30 vehicles and is well situated, being on the route of the coastal footpath.

acres) in all, is largely coastal marshland, protected to some extent by the large shingle bank along this part of the coast.

It was the first reserve acquired by the Norfolk Naturalists' Trust way back in 1926, and is noted for the large number of rare migrants nesting there, including bearded tit, bittern, avocet and black-tailed godwit. Nesting wildfowl include gadwall, shelduck and garganey.

Wildfowl numbers are impressive early in the year and wigeon, teal, mallard, shoveler and pintail are numerous. During May, Arctic-bound waders passing through include Kentish plover and Temmink's stint. Rough weather in the autumn drives many seabirds to the inshore area. All four skuas may be seen, along with guillemots, razorbills, kittiwakes, gannets, shearwaters and petrels.

Access to the East Bank is unrestricted. There is a large car park just in front of the Visitor Centre, which is clearly visible from the coast road. The Centre is quite impressively set out and contains display material outlining the history and conservation aspects of the Cley area. It is open during the season, except on Mondays. Visitors wishing to use the bird hides on the reserve should enquire at the Centre between 10 am and 5 pm; otherwise permits may be obtained in advance from the Warden, Watcher's Cottage, Cley-next-the-Sea, Holt, Norfolk (Tel: Cley 380).

27. TITCHWELL MARSH (NORFOLK) RSPB RESERVE

This RSPB reserve comprises 200 hectares (500 acres) of shingle beach, reedbed, saltmarsh and sand dunes. The land was reclaimed for agriculture in the 1780s but has recently reverted to its former state. It lies 10 km (6 miles) east of Hunstanton on the A149 Brancaster road between Titchwell and Thornham.

Bird species include bittern, common and little tern, scoter, Brent geese, goldeneye and eider duck. The area supports a rich flora, with sea lavender, sea aster, sea pink, sea arrow-grass and glasswort being particularly common. The beach strandline is most interesting and over 20 different types of shell are to be found, together with whelk cases, hornwrack and other cast-offs. At low tide, the ancient forest is visible.

An all-year car park exists at the reserve together with a small picnic area near the Information Centre. The Centre opens on most days in the summer. Toilets are available at Brancaster Beach. Access is along the sea-wall footpath to the three hides. There is no visiting charge and further information is available from the Warden, Three Horseshoes Cottage, Titchwell, King's Lynn PE31 8BB.

28. THE WASH (LINCOLNSHIRE/ NORFOLK)

The Wash is one of the biggest estuary areas in

Britain and in fact is composed of the estuaries of four rivers – the Ouse, Nene, Trent and Welland. It also boasts the largest area of saltmarsh, stretching from Gibraltar Point in the north round to the Snettisham RSPB reserve on the Norfolk coastline. A region near Gedney Drove End on the Norfolk/Lincolnshire border is used as an RAF bombing range occasionally, and is therefore prone to major disturbance.

Wildfowl and wader roosts exist all along the coastline, making the Wash a site of major importance for birds. No fewer than 12 species occur in internationally important numbers – in particular grey plover, knot and shelduck. Autumn is an excellent time to observe migrant waders, many of which move on to winter elsewhere.

Breeding birds include shelduck, redshank, ringed plover and both common and little terns. There is also a substantial black-headed gull nesting colony.

Like several other coastal regions, the Wash is also under threat from a number of influences. Pressure from holidaymakers is intense during summer and there is also a great deal of saltmarsh reclamation for agriculture.

29. HUMBER ESTUARY (HUMBERSIDE/ LINCOLNSHIRE)

One of the longest estuaries in Britain, the Humber extends from Goole inland to Spurn Head, a 5.5-km (3½-mile) long peninsula at the mouth of the estuary.

The two large ports at Hull and Grimsby mean that there is a steady flow of shipping along the Humber and, like other major estuaries, the area is susceptible to new developments. In particular, there are plans to reclaim a large area of inter-tidal flats known as Spurn Bight – a well-known place for several species of waders and wildfowl.

The biggest saltmarshes are at Grainthorpe and Tetney, with a few smaller ones along the northern shore. The major roosting area for wildfowl is near Broomfleet and is designated a National Wildfowl Refuge. Wader roosts are more widely distributed, but major concentrations occur at Cherry Cob Sands, Spurn, Tetney and Donna Nook.

Six wader species are of international importance. These are redshank, curlew, grey plover, dunlin, sanderling and knot. Of the wildfowl, only pink-footed goose numbers reach such significance, although there are far fewer of them nowadays than in the past. The Humber is also an important site for mallard, probably the best in Britain.

30. TETNEY MARSHES (SOUTH HUMBERSIDE) RSPB RESERVE

Sandflats, sand dunes and saltmarsh, covering over 1,240 hectares (3,100 acres), form the RSPB reserve at Tetney Marshes on the Humber Estuary. The reserve is approached either through Tetney Village or by taking the minor road through North Cross from the A1031 Grimsby to Mablethorpe road. High tides provide good viewing from the sea wall.

The reserve supports a little tern colony as well as redshanks and ringed plovers. During the winter months, flocks of grey and golden plover, knot, dunlin and Brent geese can all be viewed. There is no visiting charge and further information is available from the summer warden (April–August), c/o Post Office, Tetney, Grimsby, South Humberside.

31. BLACKTOFT SANDS (HUMBERSIDE) RSPB RESERVE

Some 184 hectares (460 acres) of the Upper Humber Estuary form an important habitat of tidal reedbeds fringed with saltmarsh. The reserve is reached from the A161 Goole road and a car park exists for visitors about 1 km (½ mile) east of Ousefleet.

Many waders can be seen here while the reedbeds provide homes for reed bunting, reed and sedge warblers, water rail and bearded tit; artificially-made lagoons support redshank, gadwall and shoveler; grassy areas, grasshopper warbler and short-eared owl. Winter brings merlin, hen harrier, pink-footed geese and a large number of teal.

Sea aster is plentiful in the lagoon areas and a variety of grasses thrive on the saltmarsh. The rare beetle *Dronium longiceps* and the plum red aphid are quite common here. As well as the commoner butterflies, gatekeeper, ringlet and orange-tip also occur. The rare fen, obscure and silky wainscot moths can be found in the reedbeds.

Access to the sea wall and the public hides is available throughout the year and there is a small charge for non-members. Further information can

be obtained from the warden, Hillcrest, High Street, Whitgift, Goole, Humberside DN14 8HL.

Lindisfarne (Holy Island).

32. LINDISFARNE (NORTHUMBERLAND)

Lindisfarne National Nature Reserve is located on the west coast of Northumberland, 10 km (6 miles) to the south of Berwick-upon-Tweed, off the A1 road. Some 3,200 hectares (8,000 acres) of reserve stretch from Goswick Sands in the north to Budle Bay in the south, and include part of Holy Island which is linked to the mainland by a causeway, impassable at high tide.

Saltmarshes, mudflats and sand dunes provide a variety of habitats and, in winter especially, form an important site for wildfowl and wading birds. Wintering bird species include pale-bellied Brent geese, whooper swan, dunlin, knot, goldeneye and long-tailed duck. Passing migrants include rarities such as pine grosbeak, Richard's pipit and red-breasted flycatcher. Grey seal are common.

Plants include white scurvy grass, pink thrift and blue sea aster on the saltmarsh; viper's bugloss and hound's tongue on the dunes; and bog pimpernel, early and northern marsh orchids on the slacks between the dunes.

Access to the reserve is unrestricted and at low tide cars can be left in the car park off the causeway at Holy Island.

33. FIRTH OF FORTH (FIFE/CENTRAL/ LOTHIAN)

The Firth of Forth is one of the most important areas for estuary birds in the whole of Scotland. Sandy bays occur at Aberlady, Gullane, Gosford, Musselburgh and Gramond on the southern shore, together with Largo and Pettycur to the north. In the outer Forth, there are some vitally important islands for breeding seabirds, such as the Isle of May, Bass Rock and Inchmickery.

Few saltmarshes have developed, except at Aberlady Bay which is a nature reserve. The Forth is internationally important for 14 species of birds, several of which also breed in the area. Over 50 per cent of the British population of scaup use the estuary.

The main concentrations of wading birds occur at Aberlady, Musselburgh and Longannet in the inner estuary. The same areas are used by the wildfowl population, although there is a significant roost on Largo Bay.

Unfortunately, there are a number of threats looming in the background, including industrial developments near Grangemouth, oil spillages from ships using the oil terminal and increased pollution from several sources.

33a. ABERLADY BAY (LOTHIAN)

The whole of Aberlady Bay is a designated nature reserve and consists of a mixture of wildlife habitats including saltmarsh, creeks and sand-dunes.

Breeding birds include both Arctic and common terns; the former making up the majority of the hundred pairs or so. It is a good site for breeding eider, although the nest sites are marauded each spring by a flock of up to a hundred crows which decimate the clutches of eggs. Other birds nesting on

The northern marsh-orchid (*Dactylorhiza purpurella*) which grows in profusion in dune slacks at Aberlady Bay.

the site include dunlin, redshank, lapwing and shelduck.

The reserve is an important wintering ground for pink-footed geese, and although up to 10,000 birds can be seen during late October and early November, about half of these continue southwards to winter on the Lancashire coast. Other birds spending the winter months on the reserve include teal, wigeon, scoter, long-tailed duck and both grey-lag and barnacle geese.

In the damper parts of the dunes, masses of early marsh and northern marsh orchids are in flower during summer.

34. MONTROSE BASIN (TAYSIDE)

The Montrose Basin is an important wintering area

for waders and wildfowl with over 5,000 pink-footed geese present at times. The sand and mud banks are uncovered at low tide, revealing eel grass beds.

The estuarine basin adjoins the town of Montrose and is separated from it by a railway, beneath which there is a footpath.

35. LOCH OF STRATHBEG (GRAMPIAN) RSPB RESERVE

The Loch of Strathbeg is located to the north of the A952 Peterhead/Fraserburgh road near Crimonmogate. The 920-hectare (2,300-acre) reserve includes the loch which is separated from the sea by sand dunes. Other habitats include saltmarsh, freshwater marsh, farmland and woodland.

Over 180 bird species have been recorded and include whooper swan, grey-lag and pink-footed geese, goosander, smew and wigeon. Plants include early marsh, coral root and butterfly orchids, yellow rattle, lady's smock and grass of Parnassus. Roe deer, otters and badgers are common. Fourteen species of butterfly have been recorded and include dark-green and small pearl-bordered fritillaries.

The reserve opens on certain days of the week throughout the year, but entrance is only by advance permit from the warden. There is a charge for non-members. A reception centre with car park exists, together with two hides overlooking the loch. Further details are available from the Warden, The Lythe, Crimonmogate, Lonmay, Fraserburgh AB4 4UB.

36. MORAY FIRTH AND BEAULY FIRTH (HIGHLAND/GRAMPIAN)

The inner part of this estuary is composed of sandy flats, although a saltmarsh has developed in Munlochy Bay on the northern side. Oil developments are a constant threat to an otherwise unspoilt area.

The outer estuary is sandy with sand dune developments at places like Whiteness Head and further east at Culbin Sands.

Several species of sea duck feed in the area, notably long-tailed duck, common and velvet scoter. As far as waders are concerned, Longman Bay, Whiteness and the stretch of coast between Nairn and Burghead (including Findhorn Bay) are important.

The Moray Firth is also noted for its population of divers, including the red-throated, great northern and black-throated species.

The area above Kessock Bridge is known as the Beauly Firth. It is mainly sand and mudflats with a region of saltmarsh towards the top end. It is an internationally important site for goosander and grey-lag goose. Canada geese from the Yorkshire area moult in the region.

37. CROMARTY FIRTH (HIGHLAND)

The estuary is dominated by two bays – Nigg to the north and Udale to the south, both sandy in nature. There has been and there still is a considerable amount of industrial development going on in the area, mostly oil-related. Nigg Bay seems to be a regular target and it is ironic that here up to 10,000 wigeon feed at times, together with numerous waders and ducks. Other areas preferred by birds include Udale, Alness and Dingwall Bays.

38. DORNOCH FIRTH

Extending from the beautiful town of Dornoch to the east along to where the A9 crosses the estuary at Bonar Bridge, the Dornoch is the most northerly of the larger Scottish Firths. It must also rank as one of the prettiest, bounded on both sides by steep-sided woods.

The outer estuary is lined with sand dunes with large expanses of sandflats becoming uncovered at low water. Huge numbers of sea duck have been recorded on the outer estuary, with long-tailed duck and common scoter of international importance. A variety of wading birds feed in the area.

39. LOCH FLEET (HIGHLAND)

The estuarine waters of Loch Fleet are sheltered and calm. The tidal basin extends to some 700 hectares (1,750 acres) in all, with the A9 passing close to the top end over an artificial causeway known as 'The Mound'.

Large numbers of wildfowl and waders feed on the reserve in winter, with eider, scoter and long-

Dornoch Firth, with Portmahomack on the right.

tailed duck being quite common. Crossbill and siskin breed in the adjacent pine woodlands. There is also an area of shingle and sand dune on which typical maritime plants grow.

The reserve can be viewed from the A9 or the Golspie Little Ferry road from the north. Trust members are allowed to enter the reserve, although a permit is required for the Ferry Links Wood part. Further information is available from the Scottish Trust.

40. INNER CLYDE (STRATHCLYDE)

There has been a great deal of both urban and industrial development along both sides of the es-

Barnacle geese (*Branta leucopsis*) alighting on the north bank of the Solway.

tuary during the last century. Further plans for relatively undisturbed areas pose a significant threat to the wildlife of the area. Both recreational and wildfowling activities have had a serious impact on the birdlife.

As far as the birds are concerned, one of the important feeding areas is near Erskine where most of the waders also roost. On the north side, all sections between Dumbarton Castle and Rhu are important for duck as well as waders such as oystercatcher and redshank; the number of the latter using the estuary being the only one of international importance. Up to 35,000 gulls roost on the estuary.

41. ARDMORE (STRATHCLYDE)

This 192-hectare (480-acre) Scottish Trust reserve on the north side of the Clyde Estuary consists of a substantial area of coastline. The area is noted for waders and wildfowl, in particular oystercatcher and shelduck. Large flocks of eider winter in the vicinity.

Typical maritime plants grow in the saltmarsh region and the freshwater bog on the reserve supports bog asphodel and marsh orchid. Access is off the A814 Cardross-Helensburgh road through Ardmore Farm to the car park by the shore. Permits are required for parties containing more than six people and these are available from 21 Eastwood Avenue, Giffnock, Glasgow G46 6LS.

42. SOLWAY FIRTH (CUMBRIA/DUMFRIES AND GALLOWAY)

Separating Cumbria from the Dumfries and Galloway coast of Scotland, the Solway is yet another vitally important estuary for birds on the west coast of Britain. All the Svalbard breeding population of barnacle geese winter on the North Solway, particularly around the Caerlaverock area. In late winter, the estuary supports a fifth of the British pink-footed goose population and a similar proportion of oystercatchers in autumn.

43. CAERLAVEROCK (DUMFRIES AND GALLOWAY)

This National Nature Reserve of over 5,200 hectares (13,000 acres) stretches 10 km (6 miles) along the coastline between the River Nith and Lochar Water on the Scottish shore of the Solway Firth. It can be approached from the B725 road from Dumfries or by following the B724 from Annan to Ruthwell and then taking the B725 coastal road to Backend. A car park is provided at the western end of the reserve and also opposite Tadorna (the warden's house).

Saltmarsh, known locally as merse, forms the principal habitat. Over 8,000 barnacle geese from Spitzbergen winter here; pinkfeet and grey-lag geese are also present, as are numerous ducks and waders. The reserve also has a growing colony of natterjack toad.

The Wildfowl Trust control a sanctuary area from East Park Farm, with watchtowers and hides open to the public from September to May. The warden can be contacted at Tadorna, Hollands Farm Road, Caerlaverock, Dumfries (Tel: Glencaple 275).

44. MORECAMBE BAY (LANCASHIRE/ CUMBRIA) RSPB RESERVE

Morecambe Bay is Britain's principal estuary for waders, and 2,400 hectares (6,000 acres) of grazed saltmarsh and sandflats on the eastern part of the bay between Silverdale and Hest Bank form the RSPB reserve. 20–40,000 waders can be seen here between autumn and spring and include turnstone, ringed plover, redshank, sanderling, knot, bar-tailed godwit and dunlin. Other species of interest frequently seen are goldeneye, red-breasted merganser, shelduck, pintail and peregrine.

The marshes comprise mainly red fescue grass but a rich flora is supported on the limestone cliffs to the north of the reserve.

The Bay is best viewed, especially at high tides,

from the signal box at Hest Bank off the A1505 Morecambe to Carnforth road. An information hut opens on weekends from October to March and further details can be obtained from the Warden, 1 Newcroft, Warton, Carnforth LA5 9QD. There is no visiting charge to the reserve.

45. RIBBLE ESTUARY (LANCASHIRE/ MERSEYSIDE)

Separating the towns of Lytham St Anne's and Southport, this large and important estuary is of immense importance to birds throughout the year. It is also a centre for moulting waders during late summer, especially knot and bar-tailed godwit. One feature of this estuary is that birds use most of it; the inner part edged by extensive saltmarshes at Banks, Hesketh Bank and Longton to the south; and Warton and Clifton to the north.

There are direct links with inland habitats too. The Wildfowl Trust reserve near Burscough is a refuge at various times of year for several thousand wildfowl, many of which commute to and from the Ribble estuary. Pink-footed geese graze on land between Southport and Maghull. The flock using this site in early winter represents a sizeable proportion of the world population of this species, approaching 25 per cent in some years. There are also significant numbers of wading birds using the estuary; sanderling and knot in particular.

Up to 10,000 pairs of black-headed gulls and 1,000 pairs of common terns breed on the marshes plus several wader species – notably red-shank.

46. MERSEY ESTUARY (MERSEYSIDE/ CHESHIRE)

Unfortunately, this estuary is heavily developed along its northern and southern shores and is subject to extreme pollution at times, although rapid advances have been made in recent years to eradicate the problem.

Only the upper third of the main estuary has remained relatively free from development. Saltmarshes exist near Hale on the north side, and at Ince on the southern shore, where the Manchester Ship Canal flows alongside the estuary before they join at Eastham.

Shelduck, pintail and teal numbers are of international significance, the pintail population being almost 15 per cent of the European winter flock. With the reduction in pollution over the last 20 or 30 years, both duck and wader numbers have increased dramatically. Most feed in Stanlow and Ince Banks with the saltmarsh at Frodsham being the main roosting area. Both dunlin and redshank numbers are internationally important.

47. DEE ESTUARY (MERSEYSIDE/ CHESHIRE/CLWYD)

The Dee Estuary lies between North Wales and the Wirral peninsula of Merseyside. Extensive saltmarshes exist on either side with several deep gutters making walking dangerous. Spartina has spread rapidly over the years and now colonises vast areas. At one time, the shore at Parkgate was a sandy beach, but is now an extensive saltmarsh as a result of the effect spartina has when it spreads onto open sand.

Like other west coast saltmarshes, the number of flowering plants found on the estuary is small. Glasswort has become established in several places and is gathered in large quantities at Point of Air. Sea aster, sea purslane and scurvy grass grow in profusion too.

A group of rocks just off the coast at West Kirby are excellent sites for observing wading birds at high tide. As a result, they are badly disturbed throughout most of the year. The problem is not entirely due to ornithologists as fishermen also flock to the islands, which are known as Hilbre, Little Hilbre and Little Eye.

As many as ten species of wader and two of duck are of international importance. The numbers of knot, dunlin and oystercatcher feeding on the estuary in autumn and winter are staggering. The area is also outstanding for both black-tailed and bar-tailed godwits, with over 10 per cent of the British population wintering there. Pintail and shelduck are the most common wildfowl, with the latter breeding extensively in the area.

Black-headed gulls and common terns once bred on the marshes, but are now established elsewhere – the terns preferring artificially-made rafts on lagoons in the Shotton Steelworks area. Little terns continue to breed on the Clwyd side near Talacre, but the area is susceptible to disturbance by holidaymakers, being a popular site for local caravan-site inhabitants.

48. GAYTON SANDS (CHESHIRE) RSPB RESERVE

One of Europe's most important habitats for wintering wildfowl is located at the head of the Dee Estuary. Over 2,000 hectares (5,000 acres) of mudflats and saltmarsh at Gayton Sands can be viewed from watching points at the old Baths car park, Parkgate and adjoining footpaths below Neston. The reserve is reached by taking the B5135 to Parkgate from the A540 Chester to Hoylake road and then turning right at the Boathouse Restaurant.

Pintail, shelduck, grey plover, knot, redshank, dunlin, bar-tailed godwit, water rail, hen harrier and peregrine can all be seen at this site. The saltmarsh provides homes for skylark, shelduck and oystercatcher. Reed and sedge warblers breed in the small reed bed.

There is no visiting charge and more information can be obtained from the warden at Marsh Cottage, Denhall Lane, Burton, Wirral L64 0TG.

49. POINT OF AYR (CLWYD) RSPB RESERVE

The Dee estuary is one of the most important areas in Britain for waders and wildfowl. Excellent feeding areas are uncovered at low tide and large roosts occur at high tide, especially on the shore to the left of the colliery.

This newly-acquired RSPB reserve has no formal visiting arrangements and is not wardened. It is approached by turning off the main A548 coast road between Prestatyn and Flint onto a minor no-through-road which leads to the beach at Talacre.

During the summer, the area can become terribly crowded with holidaymakers from the nearby caravan park. It is just as well that the best time to visit the reserve is in the period October–April, preferably a few hours before high water. Roosts of waders, especially oystercatchers and curlew, can be spectacular. The path along the bank towards the colliery is an excellent platform from which to watch the birds.

Oystercatchers (*Haematopus ostralegus*) in flight at Point of Ayr.

Several species of duck winter on the reserve, including pintail, wigeon, teal, scaup and scoter. During summer, the edible glasswort or marsh samphire grows in abundance on the wet mud with sea aster prominent higher up the shore. Passage migrants include greenshank, whimbrel, spotted redshank and four species of tern.

It is best to park near the holiday camp shops as the beach can become waterlogged during the high spring and autumn tides.

50. CONWY ESTUARY (GWYNEDD)

The outer estuary above the quayside at Conwy is an area of extensive sandbanks and mudflats, best viewed from the western side of the Great Orme's Head.

The area below the road-bridge between Llan-dudno Junction railway station and Glan Conwy village is dominated by spartina marsh together with a large mudflat lower down the shore. Here shelduck regularly feed together with a variety of wading birds, especially redshank and dunlin. The area is of national importance for its oystercatcher population.

51. CEFNI SALTMARSH AND   MALLTRAETH POOL (GWYNEDD)

Bordering Newborough Forest, on the southern part of the Cefni Estuary, is a marsh of about 160 hectares (400 acres), often flooded by the sea. Sea aster, sea arrow-grass, sea rush and annual seablite thrive in the sandy silt. The marsh is part of the extensive National Nature Reserve which stretches from Newborough Warren, situated several miles south-east along the shore.

The area is an important feeding place for wildfowl and waders and is most interesting in autumn and winter. Further north towards the village of Malltraeth is a shallow pool of brackish water, separated from the foreshore by an embankment

known as Malltraeth Cob. Together with the estuary itself, the pool was a source of inspiration for the late Charles Tunnicliffe, who lived nearby. Waders regularly seen feeding at the pool include lapwing, curlew and redshank. Several species of wildfowl have also been recorded, including pintail, goldeneye and shelduck. Winter visitors include both black-tailed and bar-tailed godwit.

Access is either along the cob at Malltraeth or via several footpaths leading off the A4080 near the Forestry Commission office, halfway between Newborough and Malltraeth.

Sunset on the Conwy Estuary. The mudbanks are a favourite feeding ground for gulls and waders, whose tracks are clearly visible in the mud.

important numbers: pintail, red-breasted merganser, ringed plover, teal and wigeon. A small flock of Greenland white-fronted geese feed on Cors Fochno during winter. This is a large area of raised bog extending several miles along the southern shore, behind the saltmarsh.

52. DYFI (GWYNEDD/POWYS/DYFED)

This is the largest estuary opening into Cardigan Bay and is the most important as far as birds are concerned. Wader roosts occur on Twyni Bach, situated on the National Nature Reserve of Ynyslas Dunes on the southern side of the estuary, as well as near Ynys-hir, where an RSPB reserve exists.

There are extensive areas of spartina along the southern shore. Five bird species reach nationally

53. YNYS-HIR (DYFED) RSPB RESERVE

Situated besides the Dyfi estuary in a most beautiful setting, this RSPB reserve is approached by turning right off the main A487 Machynlleth to Aberystwyth road in the village of Furnace. The entrance is well signposted.

The reserve extends to some 320 hectares (800 acres) in all and includes a wealth of habitats, including freshwater marsh, patches of conifer and sessile oak, together with a stream and waterfall.

Summer residents include common sandpiper, pied flycatcher, redstart, nightjar and three different warblers. White-fronted geese visit the estuary in winter and other recorded wildfowl include teal, wigeon, pintail and goldeneye. Since the site is close to large woodlands and extensive moorland, buzzard, sparrowhawk, merlin, hen harrier and the occasional peregrine are frequently seen.

The reserve is open throughout the year, but on certain days of the week only. There is a charge for non-members. Up-to-date information on visiting times can be obtained from the RSPB or directly from the warden at 'Cae'r Berllan', Eglwysfach, Machynlleth, Powys SY20 8TA.

The RSPB have made a film about this reserve which is available for hire. Please contact the 'film-hire' department at Sandy.

54. BURRY INLET (DYFED/WEST GLAMORGAN)

The centre of the cockle industry in South Wales, the Burry Inlet is the combined estuary of the rivers Llwchwr, Gwili, Llan, and Morlais. The dominant river is by far the Llwchwr or Loughor, and for this reason the estuary is often referred to as the Loughor Estuary.

The estuary is dangerous in many parts because of the soft mud underfoot at low tide which can trap walkers. There are also strong currents making swimming potentially dangerous. The cockle industry revolves around the village of Penclawdd. For centuries, the fishermen have braved the elements and have gone out onto the mudflats with horse and cart to bring back their cockle harvest.

Spartina grows extensively on the saltmarshes and is gradually spreading over the vast areas to the west of Penclawdd which at one time were used by the Ministry of Defence for artillery practice. The National Nature Reserve at Whitford Burrows marks the southern boundary of the estuary.

Internationally important numbers of oyster-catcher, knot and pintail feed in the area. Oyster-catchers in particular thrive and compete directly with the interests of the cockle-fishermen to the

extent that 'shoots' have been organised in the past, much to the concern of local naturalists.

55. PETERSTONE WENTLOOGE (GWENT)

The main interest of this Gwent Trust reserve on the coast between Newport and Cardiff is ornithological, with over 170 species recorded. Throughout most of the year, some 1,000 duck use the mud in the Peterstone Wentlooge area, including mallard, wigeon, scaup, pintail, gadwall, long-tailed duck, common scoter and shoveler. Shelduck breed along the seashore.

Over 4,000 dunlin regularly feed on the mud, and occasional rarities recorded include wood sandpiper, little stint, avocet and ruff. In 1975 a spotted crake was recorded and a spotted sandpiper in 1980–81.

Access is by public footpath to the sea wall from the B4239 Newport to Rumney coast road.

56. MAGOR PILL AND COLDHARBOUR PILL (GWENT)

This part of the Severn Estuary between Magor Pill and Coldharbour is managed by the Gwent Trust by agreement with the Welsh Water Authority. The reserve is part of the Seven Estuary SSSI.

Shelduck are regular feeders on the mud at Magor Pill, together with wading birds such as ringed plover, dunlin and redshank. Whimbrel are often seen during spring and autumn, since they use the estuary as a staging post during migration.

The saltmarsh supports several plants, including scurvy grass, sea wormwood and sea aster growing amongst the common spartina grass. Access is by taking the track from Magor Pill Farm, south of Whitewall Common, down past the sewage works to the sea wall.

To the west is another Gwent Trust reserve running from Goldcliff to Coldharbour Pill; to the east is Collister Pill, where the Nature Conservancy Council have a nature reserve agreement with the farmer. The marshes, which are grazed by sheep, are an important roosting area in winter for large numbers of waders.

Appendix 1: Seabird Nesting and Colony Sites

NESTING AND WINTERING SITES

Species	Nest locations
Skuas	
Great skua (*Catharacta skua*)	Open moorland close to sea.
Arctic skua (*Stercorarius parasiticus*)	Open moorland close to sea.
Gulls	
Black-headed gull (*Larus ridibundus*)	Saltmarshes, sand dunes, shingle banks, coastal lagoons.
Common gull (*Larus canus*)	Small islands, sand dunes, shingle banks, grassy and rocky slopes above sea.
Lesser black-backed gull (*Larus fuscus*)	Typically among sand dunes and on rocky islands.
Herring gull (*Larus argentus*)	Low rocky skerries, shingle banks, sand dunes, steep slopes, broad cliff-ledges, sea-stacks.
Great black-backed gull (*Larus marinus*)	The interior plateaux of islands when in colonies. Generally, when solitary, tops of stacks, small islands, and holms.
Kittiwake (*Rissa tridactyla*)	Small ledges on precipitous cliffs.
Terns	
Sandwich tern (*Sterna sandvicensis*)	Among sand dunes, on shingle banks, or among vegetation on low rocky islands.
Roseate tern (*Sterna dougallii*)	Usually on small rocky or sandy islands.
Common tern (*Sterna hirundo*)	Sand and shingle beaches, sand dunes, saltings, marine islands and islands in coastal lagoons.
Arctic tern (*Sterna paradisea*)	Low rocky skerries, sheep-grazed holms, shingle and sand banks, also open ground near the shore.
Little tern (*Sterna albifrons*)	Shingle and shell beaches, sometimes in machair.
Divers	
Red-throated diver (*Gavia stellata*)	On margin of small but deep lochs and coastal lagoons. Usually not far from sea coast where it feeds.
Black-throated diver (*Gavia arctica*)	Winters mainly along sea coasts.
Great northern diver (*Gavia immer*)	Winters along sea coasts.
Passerines	
Rock pipit (*Anthus spinoletta*)	Breeds near sea shore in rock crevices. Mudflats and sea coasts in winter (also inland).

SOME COLONY SITES IN SCOTLAND

Key

A: over 100,000 breeding pairs
B: 10,000 to 100,000 breeding pairs
C: 1,000 to 10,000 breeding pairs
D: 100 to 1,000 breeding pairs
E: 10 to 100 breeding pairs
F: 1 to 10 breeding pairs
?: number of pairs unknown
P: probably breeds but lacking proof

Map ref.	Site	Puffin	Razorbill	Guillemot	Black guillemot	Kittiwake	Shag	Cormorant	Gannet	Manx shearwater	Fulmar	Storm petrel	Leach's storm petrel
(Map 4)	**Shetland**												
1	Hermaness, Unst	B	C	B	D	C	C	E	C		B		
2	Fetlar	C	D	D	D	D	D			E	B	?	
3	Foula	B	D	C	E	C	C			F	B	D	
4	Fitful Head, southern tip of mainland	D	E	D	E	D	E	E			C		
5	Sumburgh Head, southern tip of mainland	C	D	C	E	C	D	F			D		
6	Fair Isle	B	C	C	D	B	C				B	E	
7	**Orkney** Westray and Papa Westray	D	C	B	C	B	C	F			C	D	
8	Marwick Head	D	D	B	F	B	D				C		
9	Hoy	C	D	C	F	C	E			E	C		
10	Halcro Head to Brough Ness, South Ronaldsay	C	D	C	E	C	D				C		
11	North Rona	C	C	C	E	C	E				C	?	?
12	Sule Sgeir	D	D	C	E	C			C		C	?	?
13	Flannan Island	C	C	B		C	D			E	C		?
14	St Kilda	A	C	B	E	B	D		B	?	B	?	?
15	Rhum	C	C	C	D	D	E			B	D		
16	Ailsa Craig	E	C	C	F	C	E	F	B		E		
17	Bass Rock	D	E	D		D	D	D	C		D		
18	North Sutor of Cromarty		E	D		D	E	D			D		
19	Duncansby Head south to Skirza Head, Caithness (several colonies)	D	C	C		C	D				C		
20	Sea-cliffs either side of Cape Wrath, Sutherland	B	C	B	E	D	D	E			C		

Map 4   Seabird colony sites in Scotland.

SOME COLONY SITES IN ENGLAND AND WALES

Key

A: over 100,000 breeding pairs
B: 10,000 to 100,000 breeding pairs
C: 1,000 to 10,000 breeding pairs
D: 100 to 1,000 breeding pairs
E: 10 to 100 breeding pairs
F: 1 to 10 breeding paris
?: number of pairs unknown
P: probably breeds but lacking proof

Map ref.	Site	Puffin	Razorbill	Guillemot	Black guillemot	Kittiwake	Shag	Cormorant	Gannet	Manx shearwater	Fulmar	Storm petrel	Leach's storm petrel
(Map 5)	**England**												
1	St Bee's Head, Cumbria	F	E	C	F	C		F					
2	Calf of Man	E	D	D		D	D	F		F	D		
3	Lundy Island, North Devon	E	D	C		C	E			E	E		
4	Sea cliffs between Combe Martin Bay and Woody Bay, North Devon		D	D			F	E			E		
5	Wye Rock near Tintagel, Cornwall	D	D	D									
6	St Agnes Head, Cornwall		E	E			E				D		
7	Navax Point to Godrevy Point, Cornwall		E	E			E	E			D		
8	Scilly Isles	E	D	E		D	D	E		D	E	D	
9	Scabbacombe Head, south of Brixham, South Devon		F	D		E	E	E			E		
10	Berry Head, near Brixham, South Devon		F	D		D		F			E		
11	St Aldhelm's or St Alban's Head, Dorset	E	F	D		D	E	F			E		
12	Durlston Head, Dorset	E	F	D		D	E	F			E		
13	The Needles, westerly tip of the Isle of Wight	F	F	E		F	E	D					
14	Bempton Cliffs, Yorkshire	D	D	C		B	E	E	E		D		
15	Farne Islands, Northumbria	C	D	C		C	D	D			E		
16	**Wales** Great Ormes Head, Llandudno		E	D		D	F	E			E		
17	Puffin Island	E	D	D		D	E	D			E		
18	Cemaes Bay eastwards to south of Point Lynas, North Anglesey		E	E		F	E	E			E		
19	Carmel Head, North-west Anglesey	F	F		F			E			E		
20	North Stack, Holy Island, Anglesey	F	D	C			E	F			E		
21	Bardsey Island off the south-west tip of Lleyn Peninsula		D	E		F	E			C	F	?	
22	Sea cliffs near Capel Carmel to Trwyn Talfarach, south-western tip of Lleyn Peninsula		E	E			E	E			E		
23	Trwyn Cilan to Trwyn yr Wylfa, Lleyn Peninsula		E	C		D	E	E			E		
24	St David's Head to Cemaes Head, North Pembrokeshire coast (several colonies)		E	D			E	E			E		
25	Skomer	C	C	C		C	E	F		A	E	D	
26	Skokholm	C	D	D		C				B	F	C	
27	Grassholm	F	E	E		D	E		B				
28	Caldy Island, in western part of Carmarthen Bay	F	D	D		F	E	D			E		

Map 5 Seabird colony sites in England and Wales.

WILDFOWL FOUND AT THE COAST

Species	Location
Mute swan (*Cygnus olor*)	After breeding or non-breeders, moulting and feeding in estuarine areas and brackish lagoons.
Whooper swan (*Cygnus cygnus*)	Winter in coastal areas to feed in coastal shallows and on mudflats.
Bewick's swan (*Cygnus columbianus bewickii*)	Winter feeding in coastal areas such as estuaries. Main winter feeding in flooded marshy pastures some way from coast.
Pink-footed goose (*Anser brachyrhynchus*)	Estuaries are amongst winter roosting places. In early spring larger concentrations are now feeding in coastal areas and saline pastures.
Greenland white-fronted goose (*Anser albifrons albifrons*)	Winter in reclaimed marshland and other boggy areas near coast.
Grey-lag goose (*Anser anser*)	In winter estuaries are among the roosting areas.
Canada goose (*Branta canadensis*)	The birds that breed in Yorkshire moult in Beauly Firth, Invernesshire.
Barnacle goose (*Branta leucopsis*)	Main feeding ground in winter is 'merseland'; this is sea-washed pasture.
Brent goose: dark-bellied (*Branta bernicla bernicla*) light-bellied (*Branta bernicla hrota*)	Winter on the coast where the tide dictates their feeding times. Feed on eel grass which grows in soft mud.
Common shelduck (*Tadorna tadorna*)	Spend all year in association with saline or brackish water. Often nesting in a hole in sand dunes. Dabble for food on wet muddy surfaces. Moult in areas of extensive mudflats.
Wigeon (*Anas penelope*)	Winter in estuarine areas. Feeding out on mudflats or on saltings and sea-pasture.
Green-winged teal (*Anas crecca*)	When winter feeding at coast do so on mudflats and other muddy areas.
Pintail (*Anas acuta*)	Winter feeding occurs in estuaries and other muddy areas. Also feed on seeds of saltmarsh plants sieved from driftline.
Scaup (*Aythya marila*)	A true sea duck wintering around our coast. Dives for mussels etc.
Common eider (*Somateria mollissima*)	Breeds and feeds mainly around the coasts of northern Ireland and Scotland. Diving for mussels etc.
Long-tailed duck (*Clangula hyemalis*)	Winters around the coasts of northern Britain. Dives for mussels etc.
Common scoter (*Melanitta nigra*)	Winters around the coasts of Britain. Dives for mussels etc.
Goldeneye (*Bucephala clangula*)	Winters around the coasts of Britain. Dives for mussels etc.
Red-breasted merganser (*Mergus serrator*)	Winters coastally throughout British Isles. Dives for sticklebacks and gobies etc.
Goosander (*Mergus merganser*)	Seems only to move to coastal feeding areas when lakes and rivers are frozen. Feeds in shallow water amid seaweed-covered rocks and stones. Does not dive deeply. Takes eels and sticklebacks and other small fish.

SOME WINTERING SITES FOR WILDFOWL

Key
a: over 20,000
b: 10,000 to 20,000
c: 5,000 to 10,000
d: 2,500 to 5,000
e: 1,000 to 2,500
f: 500 to 1,000
g: 250 to 500
h: 100 to 250
i: less than 100
?: occurs in small numbers

Map ref.	Site	Mute swan	Whooper swan	Bewick's swan	Pink-footed goose	White-fronted goose	Grey-lag goose	Canada goose	Barnacle goose	Brent goose	Shelduck	Mallard	Wigeon	Green-winged teal	Pintail	Scaup	Common eider	Long-tailed duck	Common scoter	Velvet scoter	Goldeneye	Red-breasted merganser	Goosander
(Map 6)	**Scotland**																						
1.	North bank of Solway, Dumfries and Galloway	?	i	i	d		h		c		g	f	f	h	f						?		
2.	Islay, Inner Hebrides, Strathclyde		i		d	i			a		i	h	g	g			e	f					
3.	Cromarty Firth, Highland	h	h		f		g				g	f	c	f							f	h	
4.	Beauly Firth, Highland	i	i		g	f					i	f	f	g	i						i	h	f
5.	Inner Moray Firth, Highland	i			g	f					g	g	e	g	h						?	i	?
6.	Ythan Estuary and vicinity, Grampion	?	h		c	d						f	f	h			e	i			i	i	
7.	Eden Estuary, Fife				h						f	h	f	g	i		g	?	e	?	?		
8.	Tay Estuary, Tayside				d		f					e		f									
9.	*Firth of Forth* Leith to Musselburgh, Lothians															b	c	h	f	?	d	i	
10.	Aberlady Bay, Lothians		i		e						h		g				g		h				
11.	Gullane Bay, Lothians																f	i	e	h		i	
(Map 7)	**Wales**																						
1.	Traeth Bach, Gwynedd	?	?								h	g	f	h	h	?	?				i	i	
2.	Dyfi Estuary, Dyfed				i						h	f	e	f	g						?	i	
3.	Burry Inlet (Loughor Estuary), Glamorgan									i	g	h	e	g	g	?	?					?	
	England																						
4.	Bridgwater Bay, Somerset										e	e	e	g	i								
5.	Tamer Estuary, Devon	i									g	g	d	g	?	?					?	?	
6.	Chesil Fleet, Dorset	f									h	?	h	d	h	i					i	?	
7.	Poole Harbour, Dorset	i					g				h	e	f	f	f		h	?			i	h	
8.	Newton Marsh (Isle of Wight), Hampshire									h	g	g	?	g	g	h					?	?	
9.	Langstone Harbour, Hampshire	?								d	e	i	e	f	i			?			i	i	
10.	Chichester Harbour, Sussex	h								c	d	h	f	g	h		?					h	?
11.	Pagham Harbour, Sussex									h	f	h		h	i		?					?	
12.	Rye Harbour, Sussex	?									g	i	i							h			
13.	Thames-side Marsh (North Kent coast)	h		?		f					h	e	f	f	h	h					?		

No.	Site																				
14.	The Swale, Kent					f			e	e	f	d	g	i						?	i	
15.	Medway Marshes and Estuary, Kent			?	?				t	e	g	c	d	t						1	?	
16.	Inner Thames, Kent	h							g	f		g	h									
17.	Leigh Marsh, Essex								e	h	h	e	h									
18.	Foulness and Maplin Sands, Essex	h							c	g	g	e	h		?		?		?		?	
19.	Hamford Water, Essex	?							e	f	h	g	g	h			?	?		?	i	
20.	Stour Estuary, Essex	h							g	e	f	d	i	f						i		
21.	The Wash, Lincolnshire			i	d				c	b	e	c	h	g	?	h	i	e	i	h	i	
22.	Humber Estuary, Humberside				f				f	d	e	f										
23.	Teesmouth, Cleveland								e	f	h	g	i							?		
24.	Lindisfarne (Holy Island), Northumbria	?	h				f		f	f	f	b	g		?	?	e	h	f	?	i	
25.	South Bank of Solway, Cumbria			c				c														
26.	Ribble Estuary, Lancashire		?	h	b	?			e	e	d	e	e						g	?	?	
27.	Dee Estuary, Cheshire								d	e	f	f	e	?					?	i	?	

Map 6 Wildfowl wintering sites in Scotland.

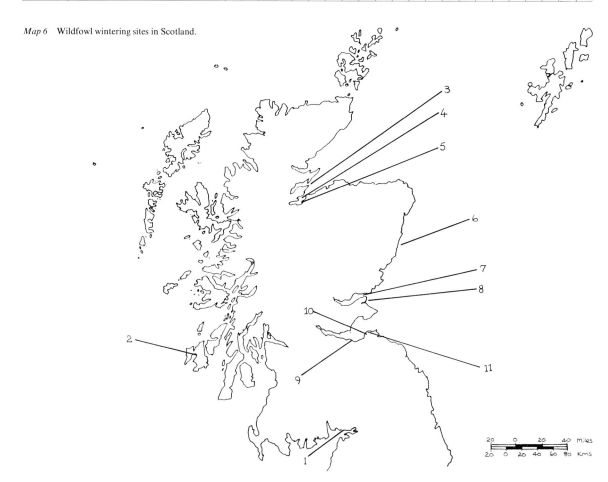

Map 7   Wildfowl wintering sites in England and Wales.

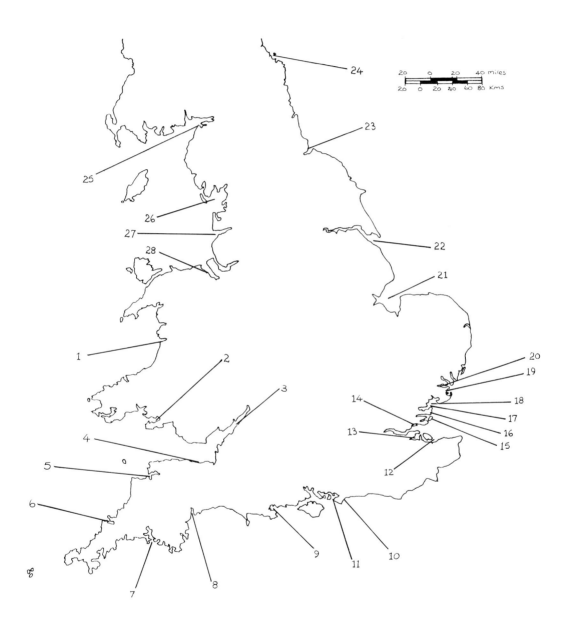

WADERS FOUND AT THE COAST

Species	Summer when breeding	Winter and on passage
Oystercatcher (*Haematopus ostralegus*)	Seashores, islands and estuaries.	Seashores, islands and estuaries.
Ringed plover (*Charadrius hiaticula*)	Beaches, dunes and saltmarshes.	Sandy and muddy shores.
Golden plover (*Pluvialis apricaria*)	Does not usually occur along coast during breeding season.	Seashores and estuaries.
Grey plover (*Pluvialis squatarola*)	Does not usually occur along coast during breeding season.	Coastal mudflats and sandy beaches.
Lapwing (*Vanellus vanellus*)	Does not usually occur along coast during breeding season.	Mudflats.
Turnstone (*Arenaria interpres*)	Does not usually occur along coast during breeding season.	Rocky and pebbly coasts.
Little stint (*Calidris minuta*)	Does not usually occur along coast during breeding season.	Seashores and estuaries.
Temminck's stint (*Calidris temminckii*)	Does not usually occur along coast during breeding season.	Occasionally saltings and estuaries.
Purple sandpiper (*Calidris maritima*)	Does not usually occur along coast during breeding season.	Rocky coasts and offshore islands.
Dunlin (*Calidris alpina*)	Occasionally breeds on saltmarshes or meadows near coast.	Seashores and estuaries.
Curlew sandpiper (*Calidris ferruginea*)	Does not usually occur along coast during breeding season.	Seashores and estuaries.
Knot (*Calidris canutus*)	Does not usually occur along coast during breeding season.	Sandy and muddy shores.
Sanderling (*Calidris alba*)	Does not usually occur along coast during breeding season.	Sandy beaches.
Ruff (*Philomachus pugnax*)	Does not usually occur along coast during breeding season.	Occasionally estuaries.
Spotted redshank (*Tringa erythropus*)	Does not usually occur along coast during breeding season.	Estuaries and mudflats.
Common redshank (*Tringa totanus*)	Saltings, occasionally breeds in meadows near coast.	Estuaries and mudflats.
Greenshank (*Tringa nebularia*)	Saltings.	Estuaries and mudflats.
Common sandpiper (*Tringa hypoleucos*)	Does not usually occur along coast during breeding season.	Estuaries when on passage.
Black-tailed godwit (*Limosa limosa*)	Occasionally breeds in meadows near coast.	Estuaries.
Bar-tailed godwit (*Limosa lapponica*)		Often seen in flocks at water's edge.
Curlew (*Numenius arquata*)	Sometimes nests in meadows near coast.	Mudflats and estuaries.
Whimbrel (*Numenius phaeopus*)	Sometimes nests in meadows near coast.	Mudflats and estuaries.
Avocet (*Recurvirostra avosetta*)	On sandbanks near shallow water.	Estuaries, mudflats, and sandbanks.

SOME WINTERING SITES FOR WADERS

Key

a: over 50,000
b: 25,000 to 50,000
c: 20,000 to 25,000
d: 10,000 to 20,000
e: 5,000 to 10,000
f: 2,500 to 5,000
g: 1,000 to 2,500
h: 500 to 1,000
i: 250 to 500
j: 100 to 250
k: up to 100
?: occurs in small numbers
×: occasional

Map ref.	Site	Oystercatcher	Ringed plover	Golden plover	Grey plover	Lapwing	Turnstone	Little stint	Purple sandpiper	Dunlin	Curlew sandpiper	Knot	Sanderling	Ruff	Spotted redshank	Common redshank	Greenshank	Black-tailed godwit	Bar-tailed godwit	Curlew	Whimbrel	Avocet
(Map 8)	**Scotland**																					
1.	Tay Estuary, Tayside	g	?	i	?	h	?	×		f	×	i	j			g	×		h	i	×	
2.	Firth of Forth (entire), Lothians	e	i	h	j	g	h		j	d	?	d	k			f	?	?	g	g	?	
3.	Cromarty Firth, Highland	g	j	?		g	k			g		g				g	?	?	h	h	×	
4.	Inner Moray Firth, Highland	f	j			h	j		?	f		f	?			g	×		g	h	×	
5.	Eden Estuary, Fife	f	j	j	j		?			f		g	?	×	×	g		j	g	?	×	
6.	North Solway, Dumfries and Galloway	d	h	f	j	e	i		?	e		e			?	f	?	?	g	f	?	
(Map 9)	**Wales**																					
1.	Dyfi Estuary, Dyfed	h	h		?	i				g	?	j				j			?	i	?	
2.	Burrey Inlet, Glamorgan	d	j	h	j	g	h		?	e		e	j		?	g	?	j	i	g	?	
3.	**England** Chittening Warth, Avon		i		?	j				f		?				j			j	i		
4.	Bridgwater Bay, Somerset	j			?	h	?			g		?			×	h		h	?	j	h	
5.	Torridge-Taw Estuary, Devon	g	i	f	k	e	j			f		j	j		?	h	?	?	?	g	?	
6.	Camel Estuary, Cornwall	i		f	?	f	j			h		?	?			j				h	?	
7.	Tamar Estuary, Devon	i	g		h	×				f		h			×	g	×	i	×	h	×	j
8.	Exe Estuary, Devon	f	i	i	i	h	i		?	e	?	i	j	?	?	h	?	h	h	h	?	?
9.	Poole Harbour, Dorset	h	j		?	i	?	×		g			×	×	×	h	×	i	k		×	
10.	Pagham Harbour, Sussex	j	j		j	j	j	×		f			×	×	×	i	×	j	?		×	
11.	Chichester Harbour, Sussex	h	j	i	h	g	j	×		c	×	i	i			g	?	h	h			
12.	The Swale, Kent	f	j	h	h	g	i			e		f	?		?	h	?	i	h	h	?	
13.	Medway Marshes and Estuary, Kent	h	h	j	h	g	k			e	?	h		?	j	g	k	i	?	g	?	
14.	Leigh Marsh, Essex	j	i		i		i			e		g				g					h	
15.	Foulness and Maplin Sands, Essex	e	i	h	i	h	j			e		e	j		?	f	?	?	g	f	?	
16.	Dengie, Essex	g	j	h	h	h	h			e		f	j			j		h	i	g	?	
17.	Blackwater Estuary, Essex	h	h	h	h	g				d		h			i	f	i	?	j		k	
18.	Colne Estuary, Essex	i	i	i	j	h	j			e	?	k	j		k	g	k	?	?	g	?	
19.	Hamford Water, Essex	i	h	g	h	g	j		?	d		j	j	?	?	g	?	j	j	g	?	
20.	Stour Estuary, Essex		j	j	i	h	j			d		h				f	?	h	?	h		

21.	The Wash, Lincolnshire	d	h	h	f		h	?		b	?	a	g		?	e	j	k	f	e	j	
22.	Humber Estuary, Humberside	g			g	i				d	?	d	i	?		g			j	g	?	
23.	Teesmouth, Cleveland	i	j	j	?	h	j	?	?	e		e	i	?		h			i			
24.	Lindisfarne (Holy Island), Northumbria	g	j	g	j	g	g	x	?	d		e	j	?	?	g	?		f	h		
25.	South Bank of Solway, Cumbria	d	g	e	j	d	i			e		d	i		?	f	k	j	f	e	?	
26.	Morecambe Bay, Lancashire	b	e	g	j	f	g			b		a	e			g			e	d		
27.	Ribble Estuary, Lancashire	f	h	g	h	f	i	?		b	?	a	e	?		f	?	h	e	h	?	
28.	Dee Estuary, Cheshire	d	g	i	i	f	i		k	b	?	b	e		?			h	f	f		

Map 8   Wader wintering sites in Scotland.

Map 9   Wader wintering sites in England and Wales.

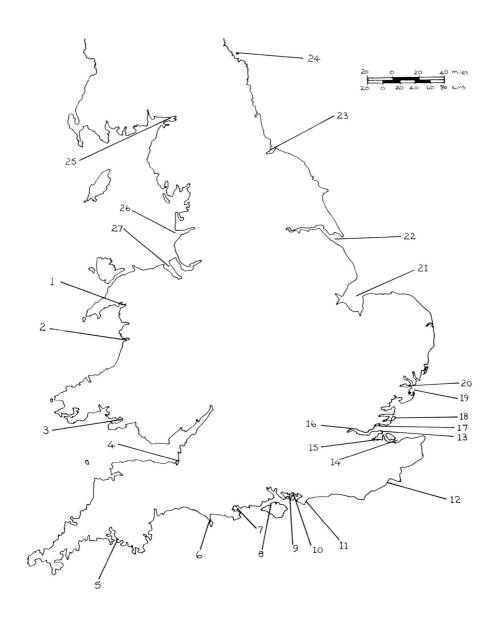

Appendix 4: Flora Checklist

Key
+ indicates an introduced species   Bi indicates a biennial
A indicates an annual   P indicates a perennial

POLYPODIACEAE (7)

Maidenhair fern (*Adiatum capillus-veneris*); rare; P.
Height 6–30 cm. In flower May/September. Fern. Damp crevices in sea-cliffs. Very local, occurs Cornwall, Devon, Dorset, Glamorgan, Cumbria, Isle of Man, W. Ireland (Clare to Donegal), Channel Is.
Sea spleenwort (*Asplenium marinum*); P.
6–30 cm. June/September. Fern. Crevices in sea-cliffs. From Isle of Wight to Cornwall; commonest on west coast northwards to Shetland; then down east coast to N. Yorks; all coasts of Ireland.
Lanceolate spleenwort (*A. billottii* [*A. obovatum*]); Very local; P.
10–30 cm. June/September. Fern. Walls, rocks and hedgebanks near the sea. Occurs in many coastal counties from West Kent to Gwynedd; also Yorkshire, Cumbria, Kintyre, Kerry, Cork, Wexford, Carlow, Wicklow, and Channel Is.

PINACEAE (11)

Maritime pine (*Pinus pinaster*); +; tree.
Up to 30 m. May/June. Planted in many coastal areas, especially in S. England. Now completely naturalised near Bournemouth.

PAEONIACEAE (15)

Wild peony (*Paeonia mascula* [*P. corallina*]); +; P.
Up to 30 cm. April/May. Deep purple-red, rarely creamy. Steep Holm (Severn Estuary) where it is naturalised.

PAPAVERACEAE (19)

Yellow horned-poppy (*Glaucium flavum*); Bi or P.
30–90 cm. June/September. Yellow. Found principally on shingle banks. Occurs in coastal areas southwards from Argyll and Kincardine; also Ireland and Channel Is.

CRUCIFERAE (21)

Wild cabbage (sea cabbage) (*Brassica oleracea*); Bi or P.
30–60 cm. May/August. Lemon-yellow. Sea cliffs in south and south-west England, also Wales.
Black mustard (*B. nigra*); A.
30–110 cm. June/August. Yellow. Sea-cliffs especially in S.W. England growing wild. Elsewhere inland where it is most likely an escape.
Isle of Man cabbage (*Rhynchosinapis monensis*); Bi.
15–30 cm June/August. Yellow. Locally common in coastal areas from N. Devon and Glamorgan north to Kintyre; also in the Isle of Man and Clyde Is.
Lundy cabbage (*R. wrightii*); P.
20–90 cm. June/August. Yellow. Peculiar to the cliffs and slopes on the east side of Lundy Island.
Sea radish (*Raphanus maritimus*); Bi or P.
20–80 cm. June/August. Yellow (white in Channel Is.). Sandy and rocky shores occurring along the driftline and on the cliffs. Southwards from Argyll and Durham; also the Hebrides and around the coasts of Ireland and the Channel Is.
Sea kale (*Crambe maritima*); P.
40–60 cm. June/August. White. Sea cliffs and rocks, along the drift-line, coastal sands and shingle. Occurs southwards from Islay and Fife; also found in Ireland.
Sea rocket (*Cakile maritima*); A.
15–45 cm. June/August. Purple, lilac or white. Along drift-line on sandy and shingly beaches. Found all round the British Is.
Hare's-ear cabbage (*Conringia orientalis*); +; A.
10–50 cm. May/July. Yellowish- or greenish-white. Occurs as a frequent casual on coastal wasteland and sea-cliffs, in many areas.
Narrow-leaved pepperwort (*Lepidium ruderale*); A or Bi.
10–30 cm. May/July. Greenish. Usually near the sea in waste places and waysides. Throughout England and especially in East Anglia. Rare in Scotland; seems absent Ireland and Channel Is.
Broad-leaved pepperwort or dittander (*L. latifolium*); P.
50–130 cm. June/July. White. Occurs in wet sand and salt-marshes. N.E. England; also southwards from Wales and Norfolk; S. Ireland.

Scurvy-grass (*Cochlearia officinalis*); Bi or P.

5–50 cm. May/August. White (rarely lilac). On cliffs and banks by the sea; also the drier brackish- and saltmarshes. Widely distributed throughout the British Isles.

Scottish scurvy-grass (*C. scotica* [*C. groenlandica*]); Bi or P.

5–15 cm. June/August. Pale mauve. Occurs locally in maritime regions of N. Scotland, the Hebrides, Orkneys and Shetlands; also Isle of Man; N and W. Ireland.

Danish scurvy-grass (*C. danica*); A.

10–20 cm January/June. Mauve or whitish. Occurs locally on both sandy and rocky shores, also on banks and walls by the sea. In coastal areas all around the British Isles.

Long-leaved scurvy-grass (*C. anglica*); Bi or P.

8–35 cm. April/June. White to pale mauve. Locally common in estuaries and on muddy shores all round British Isles with the exception of Orkney and Shetland.

Stock or gilliflower (*Matthiola incana*); rare; A or P.

30–80 cm. May/July. Purple, white, or red. Occurs but rarely on sea-cliffs locally in S. England, Isle of Wight, S. Wales and more recently in Durham.

Sea stock (*M. sinuata*); rare; Bi.

20–60 cm. June/August. Pale purple. Occurs rarely in sand dunes and sea-cliffs in Kent, N. Devon, Pembroke, Glamorgan; also in Ireland: Wexford, Clare.

RESEDACEAE (22)

Upright mignonette (*Reseda alba*); +; A or P.

30–75 cm. June/August. White. Occurs as a frequent casual in waste places particularly close to sea ports, especially in S.W. England.

TAMARICACEAE (27)

Tamarisk (*Tamarix anglica*); +; shrub.

1–3 m. July/September. Pink or white. Naturalised in a number of coastal regions from Suffolk to Cornwall, also Channel Is.

FRANKENIACEAE (28)

Sea heath (*Frankenia laevis*); P.

Up to 15 cm. July/August. Pink. On gravelly or sandy soils on landward side of saltmarshes. Found locally in coastal regions from W. Norfolk south and west to Hampshire, also the Channel Is.

CARYOPHYLLACEAE (30)

Sea campion (*Silene maritima*); P.

8–25 cm. June/August. White. On sea-cliffs, stony ground and shingle beaches. Locally abundant all round the British Isles.

Dark green mouse-ear chickweed (*Cerastium atrovirens*); A.

7.5–30 cm. May/July. White. Sandy and stony places near the sea. Locally common all round British Isles.

Little mouse-ear chickweed (*C. semidecandrum*); A.

Calcareous 1–20 cm. April/May. White. Coastal dunes, and sandy soils but by no means confined to the coast. Occurs throughout most of British Isles, not in Shetland. Chiefly coastal in Ireland.

Sea pearlwort (*Sagina maritima*); A.

2–15 cm. May/September. White. Occurs on sea-cliffs and rocks, also dune slacks. Local all round coasts of Britain.

Sea sandwort (*Honkenya peploides*); P.

10–45 cm. May/August. Greenish-white. Sandy shingle and mobile sand. Common all round British Isles.

Cliff sand-spurrey (*Spergularia rupicola*); P.

5–15 cm. June/September. Deep pink. Sea-cliffs, rocks and walls. Local and found mainly along coasts from Hampshire to Cornwall and north to Ross, Isle of Man, Inner and Outer Hebrides, Irish coast. Also coasts of Norfolk, Mid- and East Lothian, and Aberdeen.

Greater sand-spurrey (*S. media*); p.

10–38 cm. June/September. Pink with white base. Muddy and sandy saltmarshes. In areas all round the British coast.

Lesser sand-spurrey (*S. marina*); A.

10–38 cm. June/August. Pink with white base. In the drier areas of sandy and muddy saltmarshes and brackish marshes. To be found in suitable coastal regions around Britain.

Four-leaved all-seed (*Polycarpon tetraphyllum*); rare; A.

5–15 cm. June/July. White. Found locally in sandy and waste places in coastal areas of Dorset, S. Devon, and Cornwall.

Strapwort (*Corrigiola littoralis*); A.

5–25 cm. July/August. White (sometimes tipped red). On gravelly and sandy banks of pools. Very local. Near Helston in Cornwall and at Slapton Ley in S. Devon, also Channel Is.

Glabrous or Smooth rupture-wort (*Herniatria glabra*); rare; A or Bi.

6–20 cm. July/August. Greenish-white. Very local in dry sandy places including S. Devon, Suffolk, Cambridge, Norfolk, S. Lincoln and Cumbria.

Ciliate rupture-wort (*H. ciliolata*); very rare; P.

5–25 cm. July/August. Greenish-white. Maritime sands and rocks. Only recorded at Lizard Point in Cornwall and the Channel Is.

Hairy rupture-wort (*H. hirsuta*); +; A or P.

5–15 cm. July/August. Greenish-white. Naturalised on sandy ground at Christchurch in Hampshire.

Whorled knot grass (*Illecebrum verticillatum*); A.

5–20 cm. July/September. White. Grows very locally in moist sandy places. Kent, Hampshire and Cornwall.

CHENOPODIACEAE (34)

Stinking goosefoot (*Chenopodium vulvaria*); rather rare; A.

1–2.5 cm. July/September. Occurs on the landward edges of shingle beaches and saltmarshes. From Cornwall to Kent and north to Durham. Also southern Scotland, where it is an introduced species.

Glaucous goosefoot (*C. Botryodes*); A.

5 30 cm. July/August. Very local. Found on the muddy edges of

creeks and saltmarsh ditches close to the sea. Occurs from Norfolk south and west to Hampshire.

Sea beet (*Beta vulgaris* ssp. *maritima*); A, Bi or P.
30–120 cm. July/September. On seashores throughout the British Isles.

Shore orache (*Atriplex littoralis*); A.
Up to 100 cm. July/August. Usually on muddy substrates near the sea. Around the coasts of Britain. Local in Ireland and Scotland.

Hastate orache (*Atriplex hastata*); A.
Up to 100 cm. July/September. Above high tide mark on sand, shingle or mud. Common. Also found inland.

Babington's orache (*A. glabriuscula*); A.
Up to 100 cm. July/September. (Regarded as a ssp. of *A. hastata* from which it is distinguished by bracteoles which are united up to about the middle.) Above high tide mark on gravelly or sandy shores. Found around coasts of British Isles.

Iron-root or Common orache (*A. patula*); A.
Up to 100 cm. July/September. In open habitats, cultivated ground and waste places near the sea. Less common than *A. hastata*.

Frosted orache (*A. laciniata*); A.
Up to 30 cm. August/September. Sandy and gravelly shores about high tide mark. Local around coasts of Britain. Occurs on south and east coasts of Ireland.

Sea purslane (*Halimione portulacoides*); Small shrub.
Up to 80 (150) cm. July/September. Grows along edges of pools and channels on saltmarshes, usually flooded at high tide. Locally abundant in England as far north as Northumbria and Cumbria. In Scotland occurs in Wigtown, Ayr and Outer Hebrides. Ireland from Galway Bay along the south and east coasts to Down.

Herbaceous seablite (*Suaeda maritima*); A.
7–30 (60) cm. July/October. Common in saltmarshes and on seashores around the coasts of British Isles, usually occurs below high-water mark spring tides. Not recorded in Berwick, Banff, W. Sutherland or Caithness.

Shrubby seablite (*S. fruticosa*); Shrub.
40–120 cm. July/October. Local above high-water mark spring tides. Occurs on shingle banks and other well-drained situations by the sea. S. Dorset, Isle of Wight, E. and W. Kent, N. and S. Essex, E. Suffolk, E. and W. Norfolk, Lincoln and Glamorgan.

Saltwort or Prickly saltwort (*Salsola kali*); A.
Up to 60 cm. July/September. A decumbent or prostrate species found on sandy shores around the coasts of British Isles.

Glasswort or Marsh samphire (*Salicornia perennis*); P.
Up to 30 cm. August /September. Found in southern England on saltmarshes and gravelly foreshores.

Glasswort or Marsh samphire (*S. ramosissima*); A.
Erect or prostrate, 3–40 cm. August/September. On bare, firm mud in the upper part of saltmarshes. Occurs in S. and E. England, also S. Wales. Common and widespread in suitable habitats. Of uncertain distribution. A similar species occurs in S. and E. Ireland.

Glasswort or Marsh samphire (*S. europaea*); A.
Erect, (10) 15–30 (35) cm. August. Plant at first dark green, later yellowy-green, finally flushed with pink or red. Grows locally in saltmarshes on open sandy mud. Occurs on S. and W. coasts of England. Probably also in Ireland.

Glasswort or Marsh samphire (*S. obscura*); A.
Erect, 10–40 (45) cm. August/September. Plant at first of dull

glaucous green, later becoming dull yellow. Lowest of primary branches about half the length of main stem. Occurs on saltmarshes growing on bare damp mud and along the edges of channels. Found in suitable habitats along south coast, also west coast North to Cheshire, and east coast north to Lincolnshire. Elsewhere uncertain.

Glasswort or Marsh samphire (*S. pusilla*); A.
Erect, up to 25 cm (rarely prostrate). August/September. Plant yellowish-green, much branched and bushy. Later brownish- or pinkish-yellow, branches frequently tipped with bright pink. Occurs chiefly along the drift-line in the drier parts of saltmarshes. Along south coast from Dorset to Kent, also Essex and Norfolk. Carmarthen in Wales; Waterford in Ireland.

Glasswort or Marsh samphire (*Salicornia nitens*); A.
Erect, 5–25 cm. September. Lowest of primary branches less than a quarter of the length of main stem. Plant green or yellowish-green, smooth, shiny and somewhat translucent. Later becoming light browny-purple to browny-orange. Occurs in the upper parts of saltmarshes, in pans and on bare damp mud. Common in suitable places S. and E. England from Hampshire to E. Suffolk.

Glasswort or Marsh samphire (*S. fragilis*); A.
Erect, (10) 15–30 (40) cm. August/September. Lowest of primary branches usually less than a quarter of the length of main stem. Plant dull green, later dull yellowish-green. Found in the lower levels of saltmarshes on soft mud, especially along the sides of channels; below the marginal growth of Sea Purslane. Occurs from E. Suffolk down to Kent. Possibly elsewhere.

Glasswort or Marsh samphire (*S. dolichostachya*); A.
Erect, 10–40 (45) cm. July/August. A numerously-branched and bushy plant, with lowest branches equal in length to main stem. Dark green at first, later pale green or dull yellow, finally brownish. Ferile spikes often having slight purplish flush. Found at lowest level in saltmarshes, principally on muddy sand and rather firm mud. Sometimes along sides of channels. Widely distributed along coasts of Gt Britain from Lancashire to Devon, also Kent and E. Ross. S. and E. coasts of Ireland, also W. Galway.

Glasswort or Marsh samphire (*S. lutescens*); A.
Erect, (10) 15–30 (40) cm. July/August. Similar in habit to *S. dolichostachya*. Plant bright green to yellow green at first, later bright yellow. Grows on firm and dryish mud or muddy sand. Occurs on E. and S. coasts of England from S. Lincolnshire to S. Hampshire. Also Glamorganshire. Elsewhere distribution not known.

MALVACEAE (37)

Dwarf mallow (*Malva neglecta*); A or P.
15–60 cm. June/September. Pale lilac or whitish with lilac veins. Along drift-lines throughout the British Isles except Outer Hebrides and Shetlands. Common in the south, less frequent in the north.

Mallow (*M. pusilla*); +; A.
15–60 cm. June/September. Pale rose to whitish. Occurs along foreshores. An introduced species occurring locally northwards to Aberdeen.

Tree mallow (*Lavatera arborea*); Bi.
Almost tree-like. 60–300 cm. July/September. Pale rose to purple

with broad deep purple veins. Occurs up to about 155 m. Maritime rocks and waste ground near the sea. Native along S. and W. coasts of Gt Britain from Sussex to Cornwall, then northwards to Anglesey, Isle of Man and Ayr. Probably introduced on east coast where it is found as far north as Fife.

Tree mallow (*L. cretica*); A or Bi.
(Rarely prostrate) 50–150 cm. June/July. Lilac. Found along waysides and in waste places near the sea. Occurs W. Cornwall, Scilly Isles, and Jersey. Elsewhere a casual.

Marsh mallow (*Althaea officinalis*); P.
60–120 cm. August/September. Pale pink. Locally common. Upper margins of salt- and brackish-marshes, also along the sides of ditches and banks near the sea. Occurs in coastal Britain northwards to Northumbria in the east; and to Dumfries, Kirkcudbright and Arran in the west; also Jersey.

GERANIACEAE (39)

Bloody cranesbill (*Geranium sanguineum* var. *prostratum*); P.
(Stems procumbent) 10–40 cm. July/August. Bright purplish-crimson (occasionally pink: var *lancastrense*). Occurs on fixed dunes and maritime sands. Widely distributed throughout the British Isles, but seemingly absent in S.E. England and S. Ireland.

Herb robert (*G. robertianum* ssp. *maritimum*); usually Bi.
10–50 cm. May/September. Bright pink, occasionally white. Has an open procumbent rosette of leaves in the first year. Found more or less all round the British Isles. Grows on shingle beaches, also but less frequently on cliffs and walls near the sea.

Sea storksbill (*Erodium maritimum*); A.
Up to 30 cm. May/September. Pink. Chiefly near the sea. Open habitats in short dry grassland, also fixed dunes. Grows locally on S. and W. coasts from Kent to Cornwall and north to Wigtown. Also coastal areas of Norfolk, Durham and Northumbria; coasts of Ireland southwards from Down and Clare, local; Channel Isles.

Musk storksbill (*E. moschatum*); A.
Up to 60 cm. May/July. Rosy-purple. Usually robust but occasionally dwarf and almost stemless. Smells of musk. Found in waste places chiefly near the sea. Grows locally from Cornwall to Kent, S. Lancashire and the Isle of Man.

Storksbill (*E. glutinosum*); A.
Up to 25 cm. May/August. Pale pink or white. Plant slender, greyish, very hairy and sticky with sand grains adhering. Grows locally near the sea on sandy ground and dunes. Distribution not fully known, but occurs in numerous places particularly on the W. coast.

PAPILIONACEAE (51)

Small restharrow (*Ononis reclinata*); A.
4–8cm. June/July. Rose. Found by the sea in sandy soil. Occurs Berry Head, Devon; Glamorgan; Alderney and Guernsey, Channel Is. Introduced Wigtown.

Hairy medick (*Medicago polymorpha*); A.
5–60cm. May/August. Yellow. Near the sea on gravelly ground

or in sandy soil. Native in eastern and southern England; elsewhere introduced.

Spotted medick (*M. arabica*); A
10–60 cm. April/August. Yellow. Especially on sandy or gravelly soils near the sea in Britain. Now naturalised in a few localities in Ireland.

Birdsfoot fenugreek (*Trifolium ornithopodioides*); A (or P?).
2–20 cm. May/September. White and pink. Grows locally in sandy and gravelly situations mainly near the sea. Found in suitable habitats northwards to Renfrew and Fife; also S. and E. coasts of Ireland.

Clustered clover (*T. glomeratum*); (rare); A.
5–25 cm. June/August. Purplish. Found in grassy places on gravelly and sandy soils chiefly near the sea. Occurs from Cornwall and Kent to Carmarthen and Norfolk but sparingly. In Ireland, Wexford and Wicklow very local and rare.

Sea clover (*T. squamosum* [*T. maritimum*]); A.
Up to 40 cm. June/July. Pink. Occurs locally in turf close to the sea or tidal estuaries. Found in coastal areas of southern England, northwards to Carmarthen and Lincolnshire; also Gwynedd and Lancashire.

Kidney-vetch or ladies' fingers (*Anthyllis vulneraria*); P.
Erect or decumbent up to 60 cm. June/September. Yellow. Bluish-white foliage. Generally distributed throughout British Isles, but more abundant on calcareous soils and near the sea.

Hairy birdsfoot trefoil (*Lotus hispidus*); P.
3–30 (90) cm. July/August. Yellow. Found near the sea in grassy places. Occurs in Hampshire, Dorset, Devon and Cornwall, also Pembroke and Channel Is.

Slender birdsfoot trefoil (*L. angustissimus*); A.
3–30 (90) cm. July/August. Yellow. Similar to *L. hispidus* but an annual. Found near the sea in dry grassy places. Occurs Surrey, E. Kent, Sussex, Hampshire, Devon, Cornwall and Channel Is.

Wood vetch (*Vicia sylvatica* var. *condensata*); P.
Trailing, 60 cm. June/August. White with blue or purple veins. Occurs on shingle and cliffs by the sea. Scattered locally throughout British Isles.

Yellow vetch (*Vicia lutea*); A.
Prostrate 10–45 (60) cm. June/August. Pale sullied yellow. On sea cliffs and shingle. Scattered but local around the British coasts north to Kincardine and Ayr.

Sea pea (*Lathyrus japonicus* [*L. maritimus*]); P.
Creeping 30–90 cm. June/August. Purple to blue. Occurs on shingle beaches. Very local. From Lincolnshire to Dorset and Cornwall; Glamorgan; Angus; W. Ross; and Co. Kerry.

ROSACEAE (52)

Silverweed (*Potentilla anserina*); P.
Stolons up to 80 cm. June/August. Yellow. Coastal dunes occasionally shingle. By no means confined to coastal regions. Common throughout the British Isles.

Burnet rose (*Rosa piminellifolia*); shrub.
10–40 (100) cm. May/July. Creamy-white, rarely pink. Dunes and sandy heaths, especially near the sea but also inland. Rather local, occurs northwards to Caithness and the Outer Hebrides; throughout Ireland. Absent in a number of counties.

CRASSULACEAE (54)

Rose-root or midsummer-men (*Sedum rosea*); P.
15–30 cm. May/August. Greenish-yellow. Occurs in sea-cliff crevices (also mountain rocks inland to about 1250 m) S. and N. Wales, then from Lancashire and Yorkshire northwards; Isle of Man; Ireland.

English stonecrop (*S. anglicum*); evergreen P.
2–5 cm. June/July (August). Petals white tinged, pink on back. Rocks both coastal and inland, also dunes and shingle. Common in the west, from the Hebrides southwards; far less so on south and east coasts. Occurs throughout Ireland.

Wall-pepper (*S. acre*); evergreen P.
2–10 cm. June/July. Bright yellow. Dunes, shingle and walls. Not confined to coastal regions. Occurs throughout British Isles except Shetland. Common.

ELAEAGNACEAE (65)

Sea buckthorn (*Hippophaë rhamnoides*); shrub.
1–3 m. March/April. Greenish (very small). Mainly on fixed dunes, occasionally sea-cliffs. From Yorkshire south to Sussex, also Lancashire. Often planted in other coastal areas.

ONAGRACEAE (66)

Evening primrose (*Oenothera biennis*); +; Bi.
50–100 cm. June/September. Yellow. Naturalised locally. Occurs dunes, roadsides and wasteplaces; not necessarily coastal. In Britain northwards to Perth and Angus; also Ireland and Channel Is.

Evening primrose (*O. erythrosepala*); +; Bi.
50–100 cm. June/September. Yellow. Longer hairs have red bulbous bases (not present in *O. biennis*). Naturalised locally in dunes, waste places, roadsides; also inland. Occurs in many areas throughout Great Britain; Ireland; and Channel Is.

Evening primrose (*O. stricta*); +; A or Bi.
50–90 cm. June/September. Yellow at first later turns red. Introduced in dunes. Now established localities as far north as Selkirk. More especially in S.W. England and Channel Is.

Small-flowered evening primrose (*O. parviflora*); +; Bi or P.
10–80 cm. June/September. Yellow. An introduced species, now established on dunes in S.W. England.

UMBELLIFERAE (75)

Sea holly (*Eryngium maritimum*); P.
30–60 cm. July/August. Bluish. Sandy and shingly shores, dunes. Around the British coast north to Shetland.

Alexanders (*Smyrnium olusatrum*); +; Bi.
50–150 cm. April/June. Yellow-green. Widely naturalised in hedges, wasteplaces and on cliffs, especially near the sea. More common in southern counties but extends northwards to Banff and Dunbarton; also throughout Ireland where it is somewhat local away from coast.

Smallest hare's-ear (*Bupleurum tenuissimum*); A.
15–50 cm. July/September (flowers 1 mm). Yellow. In salt-marshes and waste places, usually near the coast. Occurs from Kent and Cornwall north to Durham and Lancashire. Local becoming rare northwards.

Wild celery (*Apium graveolens*); Bi.
30–60 cm. June/August. Greenish-white. Alongside rivers and ditches especially near sea. Local, occurring principally in coastal counties; absent in extreme north of Scotland; occurs locally around the entire coast of Ireland.

Rock samphire (*Crithmum maritimum*); P.
15–30 cm. June/August. White (about 2 mm). Usually on sea-cliffs and rocks, occasionally shingle or sand. Occurs from Suffolk and Kent westwards to Cornwall and northwards to Ayr; also south coasts of Lewis and Islay; around coasts of Ireland but local in north.

Fennel (*Foeniculum vulgare*); P.
60–130 cm. July/October. Yellow. Probably native on sea-cliffs. Introduced or as a casual inland. Occurs around the coasts of England and Wales, southwards from Norfolk and Denbigh. Further north introduced.

Lovage (*Ligusticum scoticum*); P.
15–90 cm. July. Greenish-white, sometimes tinged with pink. Occurs on rocky coasts locally from Kirkcudbright and Northumbria northwards. Also Ireland from Co. Down to Donegal; Galway.

Hog's fennel or sulphur-weed (*Peucedanum officinale*); rare; P.
60–120 cm. July/September. Yellow (about 2 mm). Occurs as a rare and local species on banks near the sea in Essex and E. Kent.

Wild carrot (*Daucus carota*); Bi.
30–100 cm. June/August. White with central flower of umbel often red or purple. Ssp. *carota* has ovate leaves; umbels strongly concave. Occurs in fields and grassy places especially near sea. Ssp. *gummifer* has leaves narrower than *carota* and umbels more or less flat. Occurs on cliffs and dunes locally and chiefly on the south coast. Throughout the British Isles except E. Sutherland and Shetland.

EUPHORBIACEAE (78)

Purple spurge (*Euphorbia peplis*); rare; A.
1 6 cm. July/September. Yellow. Rare on shingle beaches. Occurs sporadically in Devon, Cornwall and the Channel Is.

Portland spurge (*E. portlandica*); Bi or (P?).
5–40 cm. May/September. Yellow. Very local on sea sands and dunes. Occurs from Sussex along the west and south coasts northwards to Wigtown; all around the Irish coast but seemingly rare in the north-west; also Channel Is.

Sea spurge (*E. paralias*); P.
20–40 cm. July/October. Yellow. Somewhat local on sea sands and mobile dunes. Occurs on south coast and northwards to Norfolk in east and Wigtown in west; all round Irish coast but rare in north and west; also Channel Is.

POLYGONACEAE (79)

Knotgrass (*Polygonum aviculare*); A.

Erect or spreading up to 2 m. July/October. White or pink. Very common and not confined to coast. Occurs on roadsides and waste places.

Ray's knotgrass (*P. raii*); A.
10–100 cm straggling, prostrate. July/October. White or pink. Occurs on sandy shores and fine shingle above high-water mark. On south and west coasts north to the Hebrides; all round Irish coasts; possibly elsewhere.

Sea knotgrass (*P. maritimum*); very rare; P.
10–50 cm. July/October. White or pink. Occurs on sand or fine shingle just above high-water mark. Local in Channel Is. Probably extinct in England.

Curled dock (*Rumex crispus*); P.
50–100 cm. June/October. Green. Dune slacks and shingle beaches, but common on waste ground and cultivated land almost everywhere. The commonest of the British docks.

Shore dock (*Rumex rupestris*); P.
Up to 70 cm. June/August. Green. Sea-cliffs, rocky shores, dune slacks. S. Scilly, Cornwall, S. Devon, Dorset, Glamorgan and Pembroke.

SALICACEAE (89)

Creeping willow (*Salix repens* ssp. *argentea*); shrub.
30–150 cm. April/May (prostrate to erect). Dune slacks where it is sometimes dominant.

PLUMBAGINACEAE (95)

Sea lavender (*Limonium vulgare*); P.
8–30 (40) cm. July/October. Blue-purple. Muddy saltmarshes, often abundant, sometimes dominant. In Great Britain north up to Dumfries and Fife.

Lax-flowered sea lavender (*L. humile*); P.
8–30 (40) cm. July/August. Blue-purple. Muddy saltmarshes. Gt Britain north to Dumfries and Northumberland; entire Irish coast but rare in north.

Matted sea lavender (*L. bellidifolium*); P.
7–30 cm. July/August. Pale lilac. Sandy saltmarshes (drier parts). Norfolk and Suffolk coasts.

Rock sea lavender (*L. binervosum*); P.
5–30 (50) cm. July/September. Violet-blue. Maritime cliffs, rocks, also stabilised shingle. In Gt Britain north to Lincoln and Wigtown.

Thrift or Sea pink (*Armeria maritima* ssp. *maritima*); P.
5–30 cm. April/October. Rose-pink or white. Throughout British Isles, coastal marshes and pastures, maritime rocks and cliffs.

Jersey thrift (*A. armeria*); P.
20–60 cm. June/September. Deep rose. Found on stable dunes in Jersey.

PRIMULACEAE (96)

Sea milkwort or Black saltwort (*Glaux maritima*); P.
10–30 cm. June/August. Rose (procumbent or sub-erect). Locally common in grassy saltmarshes, also rock crevices and/or at the foot of cliffs by sea or in estuaries. S. coasts of Gt Britain.

GENTIANACEAE (100)

Centaury (*Centaurium pulchellum*); A.
2–3 (15) cm. June/September. Pink (no basal rosette). Common near the sea in S. and C. England. Less common Scotland: Dumfries, E. Lothian, S. Inner Hebrides; Ireland S. and E. coasts, Cork to Dublin.

Common centaury (*C. erythraea*); A.
2–50 cm. June/October. Pink (with basal rosette). Dunes. Common England and Ireland. In Scotland north to Ross and Outer Hebrides but much less common.

Shore centaury (*C. littorale*); A.
2–25 cm. July/August. Pink (with basal rosette). Somewhat local dunes and sandy places close to sea. Practically confined to coasts of Wales, N.W. and N. England, also Scotland.

Yellow-wort (*Blackstonia perfoliata*); A.
15–45 cm. June/October. Yellow (with basal rosette). Dunes and calcareous grasslands. Common in S. England then north to Lancashire and Durham; occurs Kirkcudbright; S. Ireland to Meath and Sligo; also Jersey.

BORAGINACEAE (103)

Northern shore-wort or Oyster plant (*Mertensia maritima*); P.
Up to 60 cm. June/August. Pink, then blue and pink. Shingle near sea. Very rare in south and somewhat local in the north; Anglesey, Caernarvon and Norfolk; from Northumberland and N. Lancashire to Shetland. Rare and local in Ireland occurs E. and N. coasts Wicklow to Donegal.

Viper's bugloss (*Echium vulgare*); Bi.
30–90 cm. June/September. Pinkish-purple in bud, later bright blue. Locally common on sandy shingle, sea-cliffs and dunes. Scattered distribution throughout England and Wales. Rare in Scotland, absent in north. Only native on east coast of Ireland.

CONVOLVULACEAE (104)

Sea bindweed (*Calstegia soldanella*); P.
10–60 cm. June/August. Pink or pale purple. Locally common sandy and shingly seashores, also dunes. Suitable habitats around British Isles. Becoming rare in N. Scotland and N. and W. Ireland.

SCROPHULARIACEAE (106)

Euphrasia (*Euphrasia foulaensis*); erect, rather slender; A.
2–8 (12) cm. July/August. Usually violet, sometimes white.

Coastal pastures and sea-cliffs. Outer Hebrides, also Moray to Shetland.

Euphrasia (*Euphrasia rotundifolia*), erect, robust, few branches; A. 4–10 cm. July. White and extremely villous. Very local on sea-cliffs. Outer Hebrides, Sutherland and Shetland.

Euphrasia (*E. marshalii*); erect, robust, much-branched; A. 5–15 cm. July/August. White and extremely villous. Grassy sea-cliffs, very local Mull of Galloway, Outer and Inner Hebrides, also Sutherland to Shetland.

Euphrasia (*E. occidentalis*); erect, robust with basal branches; A. 5–15 (20) cm. May/August. White. Local on grassy sea-cliffs practically throughout British Isles, much more common in south-west.

OROBANCHACEAE (107)

Carrot broomrape (*Orobanche maritima*), rare, A–P. 10–50 cm. June/July. Pale dusky yellow, veined with purple. Coastal areas where it is parasitic on Wild Carrot (*Daucus carota*) and occasionally but rarely on Buck's-horn Plantain (*Plantago coronopus*). Rare in Kent, Dorset, Devon, Cornwall and Channel Isles.

PLANTAGINACEAE (112)

Sea plantain (*Plantago maritima*); P. 2–6 (12) cm. June/August. Brownish. Saltmarshes and short coastal turf. Found along most of Britain's coast.

Buck's-horn plantain (*P. coronopus*); Bi 0.5–4 cm. May/July. Brownish. Sandy and gravelly soils also crevices in rocks, most coastal areas in Britain and Ireland.

VALERIANACEAE (118)

Red valerian (*Centranthus ruber*); +; P. 30–80 cm. June/August. Red, sometimes white. Coastal chalk cliffs, locally abundant in suitable habitats, Britain and Ireland.

COMPOSITAE (120)

Golden samphire (*Inula crithmoides*); P. 15–90 cm. July/August. Golden yellow. Saltmarshes, shingle banks, sea-cliffs and rocks. Great Britain S. and W. coast from Essex to Wigtown and Kirkcudbright; also S. and E. Ireland, Kerry to Dublin.

Sea aster (*Aster tripolium*); P. 15–100 cm. July/October. Blue-purple or whitish. Common all around Britain on saltmarshes, also on sea-cliffs and rocks. In N. England and Scotland seems almost confined to estuaries.

Stinking mayweed (*Anthemis cotula* var. *maritima*); A. 20–60 cm. July/September. White. Coastal areas S. and C. England, rarer further north. Introduced Outer Hebrides, Orkney and Shetland.

Cotton-weed (*Otanthus maritimus*); very rare; P. 15–30 cm. August/September. Yellow. Sandy seashores and stable shingle. Now probably only occurs Jersey; and Ireland: Wexford.

Scentless mayweed. (*Tripleurospermum maritimum* ssp. *maritimum*); A–P. 10–30 cm. July/September. White. Locally common along drift-line at foot of dunes, on shingle beaches, sea-cliffs and rocks. Occurs N. Wales, N. England, Scotland and Ireland. Ssp. *inodorum* var. *salinum* occurs on sand and shingle, mainly S. England and S. Wales.

Sea wormwood (*Artemisia maritima* ssp. *maritima*); P. 20–50 cm. August/September. Pale yellow. Locally common on sea-walls and the drier parts of saltmarshes. Around coasts of Britain north to Dunbarton and N. Aberdeen, absent in extreme north. Very local on E. and W. coasts of Ireland.

Slender thistle or Seaside thistle (*Carduus tenuifloris*); A or Bi. 15–120 cm. June/August. Pale purple-red, rarely white. Locally common waysides and waste places especially near the sea. Occurs throughout lowland Britain, north to Clyde Is. and Moray; also throughout Ireland.

JUNCAGINACEAE (125)

Sea arrow-grass (*Triglochin maritima*); A or P. 15–20 cm. July/September. Rather stout herb. Grassy places on rocky shores also saltmarsh turf. Occurs in suitable habitats along entire coastline of British Isles.

ZOSTERACEAE (127)

Eel grass or Grass-wrack (*Zostera marina*); P. 20–50 (100) cm. June/September. Rhizomatus perennial with broad leaves. Found from low-water spring tides down to 4 m, on fine gravel, sand or mud. Occurs S. coasts locally, rarer northwards.

Eel grass or Grass-wrack (*Z. angustifolia*); P. 15–30 cm. June/November. Slender rhizomatus perennial with narrow leaves. Scattered in coastal areas of British Isles north to Orkney. From half-tide mark to low-tide mark but rarely down to 4 m. Occurs on estuarine mudflats and in shallow water.

Small grass-wrack (*Z. noltii*); P. (4) 6–12 (20) cm. June/October. Slender, shortly creeping, rhizomatus perennial with narrow leaves. Locally common on mud banks in creeks and estuaries; from half-tide mark to low-tide mark. Found in British Isles north to Inverness; much less common on the west coast.

POTAMOGETOACEAE (128)

Slender-leaved pondweed (*Potamogeton filiformis*); P. 15–30 (45) cm. May/August. Extensively creeping rhizomatus perennial. Coastal lakes and streams, sometimes in brackish water. Occurs Anglesey; Scotland northwards from Berwick and Ayr; Hebrides, Orkney, Shetland, N. Ireland.

RUPPIACEAE (129)

Tassel pondweed (*Ruppia spiralis*); rare; P.
30 cm or more. July/September. Peduncle 10 cm or more, leaves 1 mm wide. Brackish ditches near sea. Scattered distribution in coastal areas of England, Wales and Ireland. Scotland occurs Wigtown, Orkney and Shetland.

Tassel pondweed (*R. maritima*); P.
30 cm or more. July/September. Peduncle 0.5–5 cm, leaves 0.5 mm wide. Occurs locally in brackish ditches and saltmarsh pools. Most coasts of British Isles, north to Shetland. Becoming less frequent in north.

LILIACEAE (133)

Asparagus (*Asparagus officinalis* ssp. *prostratus*); A.
10–30 (100) cm. June/July. Male yellow, tinged red at base, Female yellow to whitish-green. Very local and rare on grassy sea-cliffs, Dorset, Cornwall, W. Gloucester and Wales. Maritime sands, Ireland: Waterford, Wexford and Wicklow; also Channel Islands.

JUNCACEAE (136)

Mud rush (*Juncus gerardii*); P.
10–30 cm. June/July. Extensive tufts or patches. Occurs abundantly in saltmarshes. upwards from just below spring-tide high-water mark. May be locally dominant. Great Britain north to Shetland, also Ireland.

Sea rush (*J. maritimus*); P.
30–100 cm. July/August. Erect, tough, densely tufted. Saltmarshes above spring-tide high-water mark, often dominant over large areas. Abundant in coastal areas of British Isles north to Inverness.

Baltic rush (*J. balticus*); P.
15–45 cm. June/August. Does not form large tufts. Dune slacks, only rarely elsewhere. E. and N. coasts of Scotland from Fife to Sutherland, also the Hebrides.

Sharp rush or Great sea rush (*J. acutus*); P.
25–150 cm. June. Tall, very robust, dense prickly tussocks. Local on sandy seashores and dune slacks. Less frequent on saltmarshes. Occurs S. and E. coasts of England to Norfolk, also Wales from Caernarvon. S.E. coast of Ireland Cork to Dublin.

CYPERACEAE (145)

Sea club-rush (*Scirpus maritimus*); P.
30–100 cm. July/August. Stout, glabrous. Rhizomes produce short tuberous tipped runners. Muddy margins of tidal rivers in shallow water; also ponds and ditches near sea. Locally abundant along coasts of Britain, north to E. Ross and Lewis.

Broad blysmus (*Blysmus compressus*); P.
10–35 cm. June/July. Glabrous, rhizome far-creeping. Marshy areas near coast, locally abundant. Scattered distribution England, also S. Scotland to Midlothian and Outer Hebrides. Wales north from Caernarvon.

Narrow blysmus or Brown club-rush (*B. rufus*); P.
10–35 cm. June/July. Similar to *compressus* but leaves involute, more or less rush-like. Locally abundant in short grass on saltmarshes. Very local in south of England. Scattered distribution in British Isles from Suffolk and Kerry (Ireland) to Shetland.

Triangular scirpus (*Schoenoplectus triquetrus*); P.
50–150 cm. August/September. Stout, glabrous. Very local in muddy banks of tidal rivers. E. Cornwall, S. Devon, Sussex, Kent; Ireland: River Shannon.

Distant sedge (*Carex distans*); P.
15–45 cm. May/June. Glabrous, densely-tufted. Coastal marshes, crevices in wet rocks. Scattered distribution in British Isles Cornwall to Shetland but rather local.

Dotted sedge (*C. punctata*); P.
15–40 cm. June/July. Glabrous, rather tufted. Coastal marshes and crevices in wet rocks by the sea. Very local. Occurs Cornwall to Hampshire; Suffolk; W. Wales; Isle of Man; Scotland: Kirkcudbright and Wigton; Ireland: Cork, Kerry, Galway and Donegal.

Long-bracted sedge (*C. extensa*); P.
20–40 cm. June/July. Glabrous, rather rigid. On damp sea-cliffs and rocks, also grassy saltmarshes. Locally common around coasts of British Isles north to Orkney.

Sand sedge (*C. arenaria*); P.
10–40 cm. June/July. Glabrous, extensively creeping. Fixed dunes and sandy places near sea. Occurs in all coastal counties.

Divided sedge (*C. divisa*); P.
(15) 30–60 (80) cm. April/June. Glabrous, creeping. Grassy places and alongside ditches near sea, also by estuaries. Locally abundant along coast from W. Gloucester south to Cornwall, east to Kent, north to Northumberland. Also Welsh coasts north from Glamorgan. Scotland: Renfrew and Angus.

Curved sedge (*C. maritima*); P.
30–60 cm. June. Glabrous, extensively creeping. Occurs locally in damp hollows on fixed dunes. Northumberland; also E. and N. coasts of Scotland from Firth of Forth northwards. Outer Hebrides, Orkney and Shetland.

GRAMINEAE (146)

Creeping fescue (*Festuca rubra*); P.
10–70 cm. May/July. More or less erect, usually stoloniferous. Grassy places in saltmarshes and dunes. Of general distribution throughout British Isles. Var. *arenaria* (far-creeping, stiff leaves, large pubescent spikelets) occurs on dunes.

Sea poa or sea manna grass (*Puccinellia maritima*); P.
10–80 cm. June/July. More or less tufted, stoloniferous. Saltmarshes and muddy estuaries. Of general distribution in coastal regions around British Isles. Common and dominant in suitable areas.

Reflexed poa (*P. distans*); P.
15–60 cm. June/July. Tufted perennial. Muddy estuaries and

saltmarshes. Generally distributed along coasts of British Isles. Less so W. coasts of Ireland. Common in places.

Borrer's manna grass (*P. fasciculata*); P.

5–50 cm. June. Tufted perennial. Very local in muddy places near sea. Occurs along coasts of S. England; Scotland: Angus; Ireland: Waterford, Wexford and Dublin.

Procumbent poa (*P. rupestris*); A or Bi.

4–40 cm. May/July. Procumbent, rarely erect. Clayey and stony seashores, also alongside brackish ditches. Occurs locally along coasts of England and Wales.

Darnel poa (*Catapodium marinum*); A.

1–13 cm. June/July. Stout, erect to decumbent or almost prostrate. Close to sea on sand, shingle and rocks. Found locally S. coasts of Britain.

Bulbous poa (*Poa bulbosa*); P.

10–30 cm April/May. Erect and tufted. Occurs on sandy shores also on limestone near coast. Very local S. England and Glamorgan.

Lop-grass (*Bromus mollis* agg.); A or P.

5–80 cm. May/July. Culms erect or decumbent, stout or slender, pubescent or glabrous. Dunes, shingle banks and sea-cliffs. Common throughout British Isles, less so in north.

Sea couch-grass (*Agropyron pungens*); P.

30–90 cm. July/September. Tufted, often glabrous, glaucous, rhizomes far-creeping. Locally dominant on dunes, also in saltmarshes. Occurs coasts and estuaries from Glamorgan south to Cornwall, west to Kent, and north to N.E. Yorkshire. Also Cheshire to Cumbria. Ireland: Dublin to E. Cork, Limerick and SE. Galway.

Sea couch-grass (*A. maritimum*); P.

30–100 cm. June/September. Erect, glaucous, glabrous leaves. Dunes on S. and E. coasts of England, also North Sea coasts.

Sand couch-grass (*A. junceiforme*); P.

25–50 cm. June/August. Glabrous, glaucous, rhizomes abundant and far-creeping. On young dunes along sandy coasts of British Isles north to Shetland.

Lyme grass (*Elymus arenarius*); P.

1–2 m. July/August. Stout, erect, glaucous. Stem creeping and freely rooting below. On dunes around coasts of British Isles, often with *Ammophila arenaria*. Local. Rare in Ireland.

Squirrel-tail grass (*Hordeum marinum*); A.

15–35 cm. June. Glaucous, more or less decumbent. Grassy places near sea. Very local S. England and Wales. Scotland: Wigtown and Kincardine.

Grey hair grass (*Corynephorus canescens*); P.

10–30 cm. June/July. Tufted and very glaucous. Behind seaward ridge of dunes in sandy places. Occurs S. coasts of Norfolk, also Suffolk and Channel Islands. Considered not to be native in Glamorgan, S. Lancashire, Moray and W. Inverness.

Marram grass (*Ammophila arenaria*); P.

60–120 cm. July/August. Stout, erect. Extensively creeping and rooting at nodes. Occurs abundantly round the coasts of British Isles where it is often dominant on dunes.

Annual beard grass (*Polypogon monspeliensis*); A.

5–80 cm. June/July. Glabrous and more or less tufted. Occurs in damp pastures close to sea. S. England and Channel Isles.

Nitgrass (*Gastridium ventricosum*); A.

10–35 cm. June/August. Glabrous, tufted. Carboniferous limestone and sandy places usually near sea. Occurs S. England and Glamorgan.

Hare's-tail (*Lagurus ovatus*); A.

7–30 cm. June/August. Erect, downy, tufted. Sandy places in the Channel Isles. Also naturalised in a few places S. England.

Sand cat's-tail (*Phleum arenarium*); A.

3–15 (30) cm. May/June. Erect grass habit, of varying length. Locally common on dunes and in sandy fields around coasts of British Isles.

Tuberous fox-tail (*Alopecurus bulbosus*); P.

25–50 cm. June. Slender, tufted. In grassy saltmarshes. Occurs locally from E. Cornwall to Pembroke, and E. Sussex to Lincoln. Also Cheshire, S. Lancashire and N.E. Yorkshire. Scotland: Sutherland and Caithness.

Sea hard-grass (*Parapholis strigosa*); A.

15–40 cm. June/August. Slender, stems usually decumbent at base, geniculate, rarely erect. In turf on saltmarshes also waste places close to sea. Local but widely distributed around British coasts north to W. Lothian and Mull.

Lesser sea hard-grass (*P. incurva*); A.

5–8 cm. June/July. Similar to *strigosa* but much smaller. Occurs in bare places on clayey or muddy shingle, usually among taller maritime vegetation. S. England Dorset to Kent and north to Norfolk, also Bristol Channel.

Cord-grass (*Spartina maritima*); P.

10–50 cm. August/September. Stout, erect, rootstock creeping. Local on tidal mudflats, S. Devon to Lincoln, also Glamorgan.

Cord-grass (*S. × townsendii*); P.

50–130 cm. June/August. Stouter than *S. maritima*. Abundant on tidal mudflats where it is frequently planted to bind mud. Occurs S. England; planted in many other districts.

Bermuda-grass (*Cynodon dactylon*); P.

10–20 cm. August/September. Extensively creeping, stems rooting at nodes. Occurs on sandy shores. Dorset, Cornwall, N. Somerset, also Isle of Wight and Channel Isles. Elsewhere a casual.

Appendix 5: An Introduction to Fossils of the Sea Shore (By Derek Berryman, BA Hons)

Fossils are the remains of life preserved in rocks. The hard parts of animal and plants often remain in sedimentary rocks such as mudstones or siltstones. In some cases the original material may be preserved, in others the original material has been replaced by other minerals such as calcite and/or fool's gold. Sometimes just an imprint remains, just as a thumb-print can be left in plasticine.

The soft parts of an animal may on rare occasions leave an impression, and when these are found they add to the knowledge of ancient life.

Trace fossils are further evidence of life, but are not parts of living things. Examples are footprints and tracks, burrows and animal droppings.

Fossils can be found at many places around the coast, and the *Directory of British Fossiliferous Localities* in your local library will tell you what you are likely to find and where.

The coasts of Southern and Eastern England contain sites where attractive ammonites and belemnites abound; the best parts being Dorset, the Isle of Wight, and the Yorkshire coasts between Whitby and Bridlington.

Rocky seashores with cliffs provide the most profitable hunting grounds. As the cliff erodes more fossils are exposed and brought onto the beach. This makes early spring, after the winter weather, the best time of year for fossil hunting. The natural barrier of cliffs creates danger from two sources: falling rocks and rising tides. Check the tide times and go onto the shore when the tide is going out. It is very easy to forget the time when you are absorbed in collecting.

known as 'St Hilda's Serpents'. The creatures were probably similar to the present day *Nautilus*. Ammonites existed in Jurassic and Cretaceous geological periods, 165 to 65 million years ago. The shells have compartments, and by filling these with gas the depth at which they floated could be controlled. A specimen may often be found in section in a worn pebble, and the internal partitions can then be clearly seen. The fossils vary in size from 3 to 30 cm (1–12 in) in diameter, although larger ones are occasionally found.

The ammonite family is very large, and evolved rapidly. The different species give a good method of measuring the age of the rocks in which they are found. The general rule is the smaller they are the older they are: size increasing with evolution. Judgement cannot be made on one example; you may easily find an immature specimen of a relatively young species.

BELEMNITES FIG. 17 (1)

These are bullet-shaped fossils, and are the hard parts of cuttlefish-like creatures which existed at the same time as the ammonites. There are many families, and again they are found in many sizes, ranging from 3 to 15 cm (1–6 in) in length. Sometimes the faint impression of the soft parts is to be seen in the rock where the belemnite is embedded. It is worthwhile collecting the slab and not removing the fossil. Belemnites are very brittle and are rarely removed in one piece.

AMMONITES FIG. 17 (2)

These are coiled shells, and in the Whitby area are

SHELLFISH FIG. 17 (3)

The third common type of fossil is shellfish. These

Fig. 17   1. Belemnites. 2. Ammonites. 3. Sea shells.

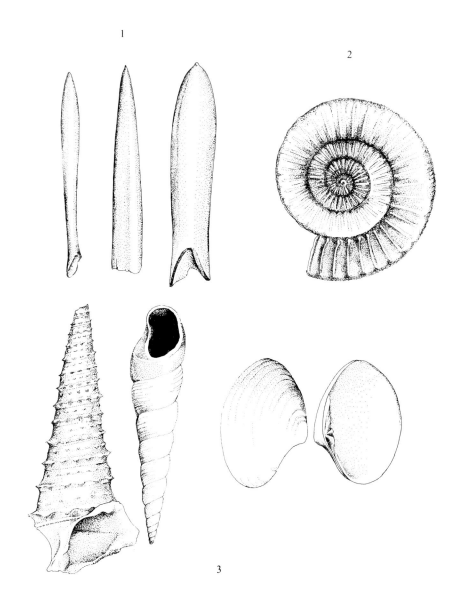

are easily recognised as oyster shells, coiled turret shells, cockles and mussels.

COLLECTING FOSSILS

The equipment needed for collecting fossils is fairly basic – newspaper and polythene bags for each sample, cold chisels and a trowel for digging-out a specimen, and a brush for cleaning away loose dirt so that you can examine the rock properly. A geological hammer and a pair of goggles are a good investment, a rucksack to carry off your haul, and don't forget your wellies!

When fossils occur in soft crumbly rock they are easily removed, but hard rocks demand the use of hammer and chisel. Always chisel away from the specimen; it is often better to finish the removal at home. Wrap each fossil in paper to stop it rubbing against others.

Fossils are often found by splitting open layered rocks and nodules. Some nodules are 'ironstone' and very hard – these contain few fossils; others may contain good specimens of ammonites.

Removal of fossils from mudstones or shales may be achieved by drying-out in an oven at moderate heat for a few hours. The crumbs can then be washed away, leaving the fossils behind. Excess rock can be removed with fine instruments, and pressure at right-angles to the fossil will often remove small flakes of rock without damage. Specimens which break can be repaired with glue.

PRESERVATION

Fossils found on the seashore contain salt which crystallises, breaking down the surface of the specimen. Soak the fossil in water for a day or so and then rinse thoroughly. You may be fortunate enough to find a fossil covered in iron pyrites (fool's gold); this may break up due to bacterial action and these should be soaked in water to which a small amount of disinfectant has been added.

When the fossils have dried the application of varnish will enhance and preserve them.

Fossils should be labelled, giving the site and date of collection, the name if possible, the age, and if from a cliff section the depth from the cliff-top, or the name of the rock bed (which may be found in local geological guides). Labels may be painted on the specimen with liquid paper if you do not wish to use ordinary adhesive labels. Boxes, drawers and shelves for keeping the specimens will no doubt be acquired by those who enjoy the thrill of fossil hunting.

IDENTIFICATION

There are many reference books available to aid general identification, local museums have displays which will help, and the opinion of curators may be sought.

Examples new to science are still being found and if you cannot easily identify a fossil take it to an expert. It may be a rarity!

Appendix 6: Seaweeds Chart

NAME OF SPECIES	Specific coastline	Specific season	Very common – common	Fairly common	Locally common	Uncommon	Rare	All shore zones	Upper (not upper extremity)	Middle	Lower	Sub-littoral	Pools	Rock pools	Deep shady rock pools	Shallow water	Low saline pools	Brackish pools and water	Semi freshwater	Rocky coasts	Especially sheltered coasts	Rocks or stones	Exposed coasts	Under ledges	Derelict boat hulls, piers and other constructions	On larger seaweeds	Among or under seaweeds	Sewage outlets	Muddy sandflats	Estuarine mudbanks	Freshwater outflows	Estuaries/lagoons/Creeks etc.	Up to (cm)
GREEN SEAWEEDS (*Chlorophyceae*)																																		
Bryopsis hypnoides	West				×					×					×							×				×								110
B. plumosa			×							×					×							×				×	×							10
Chaetomorpha aerea				×					×	×								×														×		30
C. melagonium (hog's bristle)						×				×	×		×																			×		15
Cladophora glaucescens						×				×	×		×																			×		10
C. rupestris			×							×								×		×		×	×				×					×		15
Codium fragile (fragilis)		Summer	×							×											×	×												45
C. tomentosum			×							×	×								×			×												30
Enteromorpha clathrata			×						×	×			×					×				×					×							25
E. compressa			×					×					×				×					×						×						60
E. intestinalis			×					×	×				×					×				×					×			×	×			60
E. linza			×					×					×									×					×						×	50
Prasiola stipitata				×					×													×												2.5
Ulva lactuca (sea lettuce)			×					×	+				×									×							×	×			×	45
BROWN SEAWEEDS (*Phaeophyceae*)																																		
Alaria esculenta (brown ribweed, bladderlocks, and henware)				×							×										×		×											90
Ascophyllum nodosum (egg or knotted wrack)			×							×											×	×	×											90
Aspercoccus bullosus				×						×	×											×					×							30
A. fistulosus		Summer	×							×	×		×									×												60
Bifurcaria rotunda (tuberculata)	South and South West			×						×			×									×	×											60
Chorda filum (bootlace weed, sea lace, also mermaid's tresses)			×								×	×				×		×			×	×												800
Chordaria flagelliformis		Summer	×							×			×									×												45
Cladostephus verticillatus			×							×												×				×								15
C. spongiosus			×							×												×												7
Cutleria multifida						×					×	×										×												30
Cystoseira fibrosa	South West			×						×						×						×												90
C. tamariscifolia (ericoides) (rainbow bladderweed)	South and West			×						×			×									×												60
Desmarestia aculeata					×					×												×												200
D. ligulata	South and West	Spring and Summer			×					×			×									×												200
D. viridis					×					×												×												200
Desmotrichum undulatum (Punctaria tenuissima)			×							×												×				×								30
Dictyopteris membranacea	South					×				×												×												30

Name of species	Specific coastline	Specific season	Very common – common	Fairly common	Locally common	Uncommon	Rare	All shore zones	Upper (not upper extremity)	Middle	Lower	Sub-littoral	Pools	Rock pools	Deep shady rock pools	Shallow water	Low saline pools	Brackish pools and water	Semi freshwater	Rocky coasts	Especially sheltered coasts	Rocks or stones	Exposed coasts	Under ledges	Derelict boat hulls, piers and other constructions	On larger seaweeds	Among or under seaweeds	Sewage outlets	Muddy sandflats	Estuarine mudbanks	Freshwater outflows	Estuaries/lagoons/Creeks etc.	Up to (cm)
Dictyota dichotoma	South and West; North and East		×				×			×												×											30
Ectocarpus viridis			×					×														×				× (also on shells etc)							25
E. confervoides				×						×	×											×				×							30
Elachista flaccida					×					×	×															×							25
E. fucicola			×							×																×							25
Eudesme virescens (*Castagnea virescens*)		Summer			×					+			×									× (especially on sandy coasts)											60
Fucus ceranoides (horned wrack)			×							×							×				×									×	×	90
F. serratus (serrated wrack, saw wrack)			×								+	×										×											90
F. spiralis (flat wrack, twisted wrack)			×						×	×												×											39
F. vesiculosus (bladder wrack, cutweed)			×							×											×	×									×	90
Halidrys siliquosa (pod-weed)	North					×			?	×					×							×											90
Halopteris filicina							×			×												×					×						10
H. scoparia (*Stypocaulon scoparium*)	South and North				×					×												×											15
Himanthalia elongata (*lorea*) (thong weed, sea thong)					×					×	×											×											120
Laminaria digitata (sea tangle, oarweed)			×								×	×			×							×	×										300
L. hyperborea (*cloustoni*)			×								×	×			×							×	×										200
L. saccharina (sea belt, sugary wrack)			×							×	×	×			×							×											300
Leathesia difformis (looks like a lump of yellowish-brown jelly)		Spring and Summer	·							×	×		(sometimes)									×					×						5
Litosiphon pusillus		Summer	×								×	×															×						10
Mesogloia vermiculata		Summer	×						×	×					×						×	×											7
Padina pavonia (peacock's tail or fan)	South and South West	Summer			×					×												×											15
Pelvetia canaliculata (channelled wrack)			×						× (also splash zone)													×										×	15
Petalonia fascia (*Phyllitis*)		Winter		×						×	×		×									×											30
Punctaria latifolia			×							×	×		×									×					×						30
P. plantaginea		Winter / Spring / Summer	×							×	×		×									×					×						30
Saccorhiza polychides (*bulbosa*)			×								×	×										×											490
Sauvageaugloia griffithsiana (*Mesogloia griffithsiana*)			×							+			×									×											30
Scytosiphon lomentaria (*lomentarius*)			×							×			×									×					×						45
									(also on limpet shells and eel grass)																								
Sphacelaria cirrhosa			×							×												×					×						2.5
Sphaerotrichia divaricata (*Chordaria divvaricata*)		Summer					×			×			×									×											45
Sporochnus pedunculatus		Summer				×	?				×											×											45
Stilophora rhizodes		Summer				×	?			×												×					×				×		60
Taonia atomaria	South and East; North					×	×			×												×											30
RED SEAWEEDS (*Rhodophyceae*)																																	
Ahnfeltia plicata			×							×	×		×									×	×										30
Apoglossum ruscifolium					×					×	×			×																			10
Bostrychia scorpioides					×			(grows among lower stems of plants at edge of salt marshes e.g. Sea Purslane, Common Sea-blite)																									10
Brongniartella byssoides		Summer			×					×												×											30
Calliblepharis ciliata	South; elsewhere		×				×						× (in areas too deep to reach by wading)																				30
C. lanceolata	South and West					×				×			× (on holdfasts of *Laminaria* spp.)																				20
Callophyllis laciniata					×					×	×											×				×							7
									(usually among spp. of *Laminaria*)																								

Column groups: **COAST AND SEASON** (Specific coastline, Specific season) · **STATUS** (Very common – common, Fairly common, Locally common, Uncommon, Rare) · **SHORE ZONE** (All shore zones, Upper (not upper extremity), Middle, Lower, Sub-littoral) · **WHERE FOUND** (Pools, Rock pools, Deep shady rock pools, Shallow water, Low saline pools, Brackish pools and water, Semi freshwater, Rocky coasts, Especially sheltered coasts, Rocks or stones, Exposed coasts, Under edges, Derelict boat hulls, piers and other constructions, On larger seaweeds, Among or under seaweeds, Sewage outlets, Muddy sandflats, Estuarine mudbanks, Freshwater outflows, Estuaries/lagoons/Creeks etc.) · **LENGTH** (Up to (cm))

Name of species	Specific coastline	Specific season	Very common – common	Fairly common	Locally common	Uncommon	Rare	All shore zones	Upper (not upper extremity)	Middle	Lower	Sub-littoral	Pools	Rock pools	Deep shady rock pools	Shallow water	Low saline pools	Brackish pools and water	Semi freshwater	Rocky coasts	Especially sheltered coasts	Rocks or stones	Exposed coasts	Under edges	Derelict boat hulls, piers and other constructions	On larger seaweeds	Among or under seaweeds	Sewage outlets	Muddy sandflats	Estuarine mudbanks	Freshwater outflows	Estuaries/lagoons/Creeks etc.	Up to (cm)
Catenella repens (opuntia)			×							×			(in cracks)								×	×					×							1.2
Ceramium acanthonotum			×								×	×										×												10
C. rubrum			×								×		×									×					×							30
Champia parvula	South and West	Summer				×					×	×			× (fringes)													×							30
Chondria dasyphylla		Summer	×								×		×																					30
C. tenuissima		Summer				×					×		×																					30
Chondrus crispus (carragheen)			×							×			× (grows in many places but NOT on mud)																				15
Chylocladia verticillata (Chylocladia kaliformis)		Summer		×							×	×	×										× also on shells etc											30
Corallina officinalis (coralline, coral weed)			×								×	×	× (especially if shady)										× also on shells											15
Cryptopleura ramosa			× (more so in south)								×	×			× (at times)											× also on species of Laminaria							20
Cystoclonium purpureum		Summer	×								×	×	×									×												45
Delesseria sanguinea			×									×				×										× at times on spp. of Laminaria							25
Dilsea carnosa (edulis)			×								×	×										×												38
Dumontia incrassata		Spring Summer	×								×		×										× also on shells											30
Furcellaria fastigiata				×							+		× (sandy)									×											20
Gastroclonium ovatum (Chylocladia ovatus)			×								×	×										×				×							20
Gelidium corneum				×							+				× (fringes)							×											15
G. latifolium					×						×											×											7
Gigartina stellata (carragheen moss)			×								×	×										×					×						20
Gracilaria verrucosa (Gracilaria confervoides)					×						×											×							×				45
Grateloupia filicina	South ...... × / elsewhere ...... ×					×	×			×											×										×		12
Griffithsia corallinoides		Spring and Summer			×						×		×																				20
G. flosculosa			×								×		×											×									20
Gymnogongrus griffithsiae						×					×											×											7
G. norvegicus						×					×											×											7
Halarachnion ligulatum	South Summer ...... × / elsewhere ...... ×				×		×			×											×											30
Halopitys incurvus	South				×						×											×											25
Halurus equisetifolius		Spring Summer			×			× .... ×			×											×											20
Helminthora divaricata		Summer			×						×											×					× (also on shells)						20
Heterosiphonia plumosa			×										× (in deep water below lower zone)								×					×						30
Hildebrandia spp.			×							×					× (in shade)							× (a flat encrustation)											?
Hypoglossum woodwardii				×							×	×	×									×					× (on their stalks)						20
Jania rubens (Corallina rubens)					×						×		×																				5
Laurencia hybrida (L. caespitosa)			×								+		× (at bottom or lower sides of shallow pools)									× (also on shells)											15
L. obtusa		Summer	×								+	×	×										×											15
L. pinnatifida (pepper dulse)			×								×	×	×									×											10
Lithophyllum incrustans			×								×	×	× (also on shells)									× (a thin encrustation)											?
Lithothamnion spp.			×									×	× (also on shells)									× (thin encrustation)											?
Lomentaria articulata			×								×		× (also on shells)									×				×							20
L. clavellosa		Summer	×								×	×	×									×				×							30
Membranoptera alata			×								×	×	×									×				×							20
Myriogramme bonnemaisonii						×					×											×				× (especially Laminaria spp.)							10
Nemalion elminthoides		Summer			×						×		×									× (also on shells)											25
N. multifidum		Summer		×							×		×									× (also on shells)											25
Nitophyllum punctatum				×							×	×				×											×						50
Odonthalia dentata	North		×								×		×									× (also on shells)											25
Phycodrys rubens (seaweed oak)			×								×					×										× (on stalks of Laminaria spp.)							25
Phyllophora membranifolia			×								×		×																				25
P. palmettoides	South ...... × / elsewhere ...... ×					×	×			×											×											10
P. rubens (epiphylla)					×						×	×				×						×				×							15
Plocamium coccineum			×								×	×	×									×											20
Plumaria elegans			×								×	×										× (in shaded sites)											10

NAME OF SPECIES	Specific coastline	Specific season	Very common – common	Fairly common	Locally common	Uncommon	Rare	All shore zones	Upper (not upper extremity)	Middle	Lower	Sub-littoral	Pools	Rock pools	Deep shady rock pools	Shallow water	Low saline pools	Brackish pools and water	Semi freshwater	Rocky coasts	Especially sheltered coasts	Rocks or stones	Exposed coasts	Under ledges	Derelict boat hulls, piers and other constructions	On larger seaweeds	Among or under seaweeds	Sewage outlets	Muddy sandflats	Estuarine mudbanks	Freshwater outflows	Estuaries/lagoons/Creeks etc.	Up to (cm)
Polyides caprinus (*Polyides rotundus*)						×			×				× (usually sandy)									×											20
Polyneura gmelini	South and West					×				×	×											×					×						7
	North						×																										
Polysiphonia fastigiata (*P. lanosa*)	×		×							×																×							15
P. violacea	×		×								×											×				×							15
Porphyra umbilicalis (purple laver, slake, black butter)	×	Autumn Winter and Spring	×					×														× (also concrete Breakwaters)											25
Pterocladia pinnata (*P. capillacea*)							×			×	×											× (in shady places)					×						15
Ptilota plumosa	North		×										× (especially stalks of *Laminaria* spp.)														×						20
	elsewhere					×																											
Rhodochorton spp.	×		×							×												×					×						1.2
Rhodymenia palmata (dulse)	×		×							×	×											×					× (on stalks of *Fucus* and *Laminaria* spp.)						30
R. pseudapalmata (*palmetta*)	South-west					×				×	×											×					× (usually on *Laminaria* spp.)						10
Scinaia furcellata		Summer				×					×											×											25
Sphaerococcus coronopifolius	South and West				×						×											×											30
Sphondylothamnion multifidum	South					×					×		×																				20
	elsewhere						×																										

OFFICIAL ORGANISATIONS

NATURE CONSERVANCY COUNCIL
Great Britain Headquarters
19/20 Belgrave Square, London SW1X 8PY
(Telephone: 01-235 3241)

COUNTRYSIDE COMMISSION
John Dower House, Grescent Place, Cheltenham, Glos., GL50 3RA
(Telephone: 0242-21381)
There is a separate *Committee for Wales* at 8 Broad Street, Newtown, Powys SY16 2LU, which advises the Countryside Commission on its work in the Principality and helps it to carry out its functions there.

COUNTRYSIDE COMMISSION FOR SCOTLAND
Battleby, Redgorton, Perth, PH1 3EW
(Telephone: 0738-27921)

FORESTRY COMMISSION
Headquarters: 231 Corstorphine Road, Edinburgh, EH12 7AT
(Telephone: 031-334 0303)

WATER SPACE AMENITY COMMISSION
1 Queen Anne's Gate, London SW1H 9BT
(Telephone: 01-222 8111)

VOLUNTARY ORGANISATIONS

BOTANICAL SOCIETY OF THE BRITISH ISLES
c/o Department of Botany, British Museum (Natural History), Cromwell Road, London SW7 5BD

BRITISH ASSOCIATION FOR SHOOTING AND CONSERVATION (formerly WAGBI – For Shooting and Conservation)
Marford Mills, Rossett, Wrexham, Clwyd LL12 OHL
(Telephone: 0244 570881)

BRITISH NATURALISTS' ASSOCIATION
23 Oak Hill Close, Woodford Green, Essex

BRITISH TRUST FOR CONSERVATION VOLUNTEERS
Headquarters: 36 St Mary's St., Wallingford, Oxfordshire OX10 OEU
(Telephone: 0491 39766)
(10 Regional offices throughout Britain)

BRITISH TRUST FOR ORNITHOLOGY
Beech Grove, Tring, Hertfordshire HP23 5NB
(Telephone: 044-282 3461)

COMMONS, OPEN SPACES AND FOOTPATHS PRESERVATION SOCIETY
see OPEN SPACES SOCIETY

CONSERVATION SOCIETY
12A Guildford Street, Chertsey, Surrey KT16 9BQ
(Telephone: 093-28 60975)

CONSERVATION TRUST
246 London Road, Earley, Reading, Berks, RG6 1AJ
(Telephone: 0734-663650 or 663281)
Resource Centre: George Palmer School, Northumberland Avenue, Reading, Berks, RG2 OEN

COUNCIL FOR ENVIRONMENTAL CONSERVATION (CoEnCo)
Zoological Gardens, Regent's Park, London NW1 4RY
(Telephone: 01-722 7111)

COUNCIL FOR THE PROTECTION OF RURAL ENGLAND
4 Hobart Place, London SW1W OHY
(Telephone: 01-235 9481 and 4771)

The two bodies in Scotland and Wales that have aims and activities similar to those of the Council for the Protection of Rural England are:

ASSOCIATION FOR THE PROTECTION OF RURAL SCOTLAND
14A Napier Road, Edinburgh EH10 5AY
(Telephone: 031-229 1898)

COUNCIL FOR THE PROTECTION OF RURAL WALES
14 Broad Street, Welshpool, Powys, SY21 7SD
(Telephone: 0938-2525)

COUNTY NATURALISTS' TRUSTS
see NATURE CONSERVATION TRUSTS

FAUNA & FLORA PRESERVATION SOCIETY
c/o Zoological Society of London, Regent's Park, London NW1
4RY
(Telephone: 01-586 0872)

FIELD STUDIES COUNCIL
Director and Information Office:
Preston Montford, Montford Bridge,
Shrewsbury SY4 1HW
(Telephone: 0743 850674)
London Office:
62 Wilson Street, London EC2A 2BU
(Telephone: 01-247 4651)

FRIENDS OF THE EARTH
377 City Road, London EC1V 1NA
(Telephone: 01-837 0731)

GOSLINGS *see* WILDFOWL TRUST

NATIONAL TRUST
42 Queen Anne's Gate, London SW1H 9AS
(Telephone: 01-222 9251)

NATIONAL TRUST FOR SCOTLAND
5 Charlotte Square, Edinburgh, EH2 4DU
(Telephone: 031-226 5922)

NATURE CONSERVATION TRUSTS
Avon Wildlife Trust, 209 Redland Road, Bristol BS6 6YU
Beds & Hunts Naturalists' Trust, 38 Mill Street, Bedford MK40 3HD
Berks, Bucks & Oxon Naturalists' Trust, 122 Church Way, Iffley, Oxford OX4 4EG
Brecknock Naturalists' Trust, Chapel House, Llechfaen, Brecon
Cambs & Isle of Ely Naturalists' Trust, 1 Brookside, Cambridge CB2 1JF
Cheshire Conservation Trust, c/o Marbury Country Park, Northwich CW9 6AT
Cornwall Naturalists' Trust, Trendrine, Zennor, St Ives, Cornwall TR26 3BW
Cumbria Trust for Nature Conservation, Church Street, Ambleside
Derbyshire Naturalists' Trust, Estate Office, Twyford, Barrow-on-Trent DE7 1HJ
Devon Trust for Nature Conservation, 75 Queen Street, Exeter EX4 3RX
Dorset Trust for Nature Conservation, 39 Christchurch Road, Bournemouth BH1 3NS
Durham County Conservation Trust, 52 Old Elvet, Durham DH1 3HN
Essex Naturalists' Trust, Fingringhoe Wick Nature Reserve, South Green Road, Fingringhoe, Colchester CO5 7DN

Glamorgan Naturalists' Trust, The Glamorgan Nature Centre, Tondu, Bridgend CF32 OEH
Glos. Trust for Nature Conservation, Church House, Standish, Stonehouse GL10 3EU
Gwent Trust for Nature Conservation, The Shire Hall, Monmouth, Gwent NP5 3DY
Hants & Isle of Wight Naturalists' Trust, 8 Market Place, Romsey SO5 8NB
Hereford & Radnor Naturalists' Trust, Community House, 25 Castle Street, Hereford HR1 2NW
Herts & Middx Trust for Nature Conservation, Offley Place, Great Offley, Hitchin SG5 3DS
Kent Trust for Nature Conservation, 125 High Street, Rainham, Kent
Lancs Trust for Nature Conservation (with Greater Manchester and Merseyside), Dale House, Dale Head, Slaidburn, Nr Clitheroe BB7 4TS
Leics & Rutland Trust for Nature Conservation, 1 West Street, Leicester LE1 6UU
Lincs & South Humberside Trust for Nature Conservation, The Manor House, Alford LN13 9DL
Manx Nature Conservation Trust, Ballacross, Andreas, Isle of Man
Norfolk Naturalists' Trust, 72 Cathedral Close, Norwich NR1 4DF
Northants Naturalists' Trust, Lings House, Billing Lings, Northampton NN3 4BE
Northumberland Wildlife Trust, c/o Hancock Museum, Barras Bridge, Newcastle upon Tyne NE2 4PT
North Wales Naturalists' Trust, High Street, Bangor, Gwynedd LL57 INU
Notts Trust for Nature Conservation, c/o Dr. D. Parkin, Dept of Genetics, University of Nottingham, Nottingham
Scottish Wildlife Trust, 25 Johnston Terrace, Edinburgh EH1 2NH
Shropshire Conservation Trust, Bear Steps, St Alkmunds Square, Shrewsbury SY1 1UH
Somerset Trust for Nature Conservation, Fyne Court, Broomfield, Bridgwater TA5 2EQ
Staffs Nature Conservation Trust, 3A Newport Road, Stafford
Suffolk Trust for Nature Conservation, St Edmund House, Ropewalk, Ipswich IP4 1LZ
Surrey Trust for Nature Conservation, Hatchlands, East Clandon, Guildford GU4 7RP
Sussex Trust for Nature Conservation, Woods Mill, Henfield, W Sussex BN5 9SD
Warwickshire Nature Conservation Trust, 1 Northgate Street, Warwick CV344SP
West Wales Naturalists' Trust, 7 Market Street, Haverfordwest, Dyfed
Wilshire Trust for Nature Conservation, Wyndhams, St Joseph's Place, Bath Road, Devizes, Wilts SN10 1DD
Worcestershire Nature Conservation Trust, The Lodge, Beacon Lane, Rednal, Birmingham B45 9XN
Yorkshire Naturalists' Trust, 20 Castlegate, York YO1 1RP

OPEN SPACES SOCIETY (formerly COMMONS, OPEN SPACES AND FOOTPATHS PRESERVATION SOCIETY)
25A Bell Street, Henley-on Thames, Oxon RG9 2BA
(Telephone: 04912 3535)

PEOPLE'S TRUST FOR ENDANGERED SPECIES
19, Quarry Street, Guildford, Surrey, GU1 3EH
(Telephone: 0483 35671)

RAMBLERS' ASSOCIATION
1/5 Wandsworth Road, London SW8 2LJ
(Telephone: 01-582 6878)

ROYAL SOCIETY FOR NATURE CONSERVATION
The Green, Nettleham, Lincoln, LN2 2NR
(Telephone: 0522-752326)

ROYAL SOCIETY FOR THE PROTECTION OF BIRDS
Headquarters: The Lodge, Sandy, Bedfordshire, SG19 2DL
                 (Telephone: 0767-80551)
Scottish Office: 17 Regent Terrace, Edinburgh, EH7 5BN
                 (Telephone: 031-556 5624)
Welsh Office: 18 High Street, Newtown, Powys, SY16 1AA
                 (Telephone: 0686 26678)

SCOTTISH FIELD STUDIES ASSOCIATION
Cuthbertson, Provan & Strong, 34 West George Street, Glasgow
G2 1DH, OR Enochdu, Blairgowrie, Perthshire

SCOTTISH RIGHTS OF WAY SOCIETY
23 Rutland Square, Edinburgh, EH1 2BW

SCOTTISH WILDLIFE TRUST
25 Johnston Terrace, Edinburgh, H1 2NH
(Telephone: 031-226 4602)

SOCIETY FOR THE PROMOTION OF NATURE CON-
SERVATION see ROYAL SOCIETY FOR NATURE CON-
SERVATION

TREE COUNCIL
35 Belgrave Square, London SW1X 8QN
(Telephone: 01-235 8854)

WILDFOWL TRUST
Headquarters: The New Grounds, Slimbridge, Glos., GL2 7BT
(Telephone: 045-389 333)
Other Centres:
Arundel: Mill Road, Arundel, Sussex BN18 9BP
              (Telephone: 0903-883355)
Martin Mere: Burscough, Ormskirk, Lancs L40 0TA
              (Telephone: 0704-895181)
Peakirk: Peterborough, Cambridgeshire PE6 7NP
              (Telephone: 0733-252271)
Washington: Middle Barmston Farm, Washington 15, Tyne &
              Wear NE38 8LE
              (Telephone: 0632-465454)
Welney: Pintail House, Hundred Foot Bank, Welney,
              Wisbech,
              Cambridgeshire PE14 9TN
              (Telephone: 0353-860711)
Caerlaverock: Eastpark Farm, Caerlaverock, Dumfries DG1
              4RS
              (Telephone: 038-777-200)

WAGBI – For Shooting & Conservation see BRITISH ASSO-
CIATION FOR SHOOTING AND CONSERVATION

THE WOODLAND TRUST
37 Westgate, Grantham, Lincs NG31 6LL
(Telephone: 0476 74297)

WORLD WILDLIFE FUND
Panda House, 11–13 Ockford Road, Godalming, Surrey GU7
1QU
(Telephone: (04868) 20551)

YOUNG ORNITHOLOGISTS' CLUB see ROYAL SOCIETY
FOR THE PROTECTION OF BIRDS

YOUNG PEOPLE'S TRUST FOR ENDANGERED SPE-
CIES (YPTES) see PEOPLE'S TRUST FOR ENDANGERED
SPECIES

Bibliography

Angel, H., *The World of an Estuary*, Faber & Faber 1974.

Angel, H., *Life in our Estuaries*, Jarrold 1977.

Angel, H., *The Natural History of Britain and Ireland*, Michael Joseph 1981.

Barnes, R.S.K., *Estuarine Biology*, Edward Arnold 1974.

Barnes, R.S.K., *Coasts and Estuaries*, Hodder & Stoughton 1979.

Barrett, J., and Yonge, C.M., *Pocket Guide to the Seashore*, Collins 1958.

Bruun, B., *The Hamlyn Guide to Birds of Britain and N. Europe*, Hamlyn 1970.

Campbell, A.C., *The Hamlyn Guide to the Seashore and Shallow Seas of Britain*, Hamlyn 1976.

Campbell, B., *Birds of Coast and Sea*, Oxford University Press 1977.

Chapman, V.N., *Coastal Vegetation*, Pergamon 1976.

Clapham, Tutin and Warburg, *Flora of the British Isles* (second edition), Cambridge University Press 1962.

Clayton, J.M., *The Living Seashore*, Frederick Warne 1974.

Cramp, S., Bourne, W.R.P., and Saunders, D., *The Seabirds of Britain and Ireland*, Collins 1974.

Fitter, R., Fitter, A., and Blamey, M., *The Wild Flowers of Britain and N. Europe*, Collins 1974.

Flegg, J., *In Search of Birds*, Blandford Press 1983.

Gooders, J., *Where to Watch Birds*, Andre Deutsch 1967/74.

Hale, W.G., *Waders* (New Naturalist Series), Collins 1980.

Hamilton, R., and Insole, A.N., *Finding Fossils*, Kestrel Books.

Hardy, E., *A Guide to the Birds of Scotland*, Constable & Co. 1978.

Hepburn, I., *Flowers of the Coast*, Collins 1952.

Hewer, H.R., *British Seals* (New Naturalist Series), Collins 1974.

Hill, C.A., *A Guide to the Birds of the Coast* (revised by B. Campbell and others), Constable 1976.

Kerney, M.P., and Cameron, R.A.D., *Field Guide to the Snails of Britain and North West Europe*, Collins 1979.

Kirkaldy, J.F., *Fossils in Colour*, Blandford Press 1975.

Lockley, R.M., *Eric Hosking's Sea Birds*, Croom Helm Ltd 1983.

Major, A.P., *The Book of Seaweed*, Gordon and Cremonsi 1977.

Major, A.P., *Coast, Estuary and Seashore Life*, John Gifford 1973.

Mathieson, *Let's Collect Fossils*, Jarrold 1977.

Ogilvie, M.A., *The Bird-watcher's Guide to the Wetlands of Britain*, Batsford 1979.

Ogilvie, M.A., *Ducks of Britain and Europe*, T. & A.D. Poyser 1975.

Owen, M., *Wildfowl of Europe*, MacMillan London and The Wildfowl Trust 1977.

Palaeontographical Society, *Directory of British Fossiliferous Localities*.

Parslow, J., general editor, *Birdwatchers' Britain*, Pan 1983.

Phillips, R., *Wild Flowers of Britain*, Pan 1977.

Phillips, R., *Grasses, Ferns, Mosses and Lichens of Britain and Ireland*, Pan 1980.

Prater, A.J., *Estuary Birds of Britain and Ireland*, T. & A.D. Poyser 1981.

Prestt, I., *British Birds* (*Lifestyles and Habitats*), Batsford 1982.

Ranwell, D.S., *The Ecology of Saltmarshes and Sand Dunes*, Chapman & Hall 1972.

Reader's Digest Association Ltd, *Illustrated Guide to Britain's Coast*, Drive Publications Ltd 1984.

Soothill, E., and Soothill, R., *Wading Birds of the World*, Blandford Press 1982.

Soothill, E., and Thomas, M.J., *Nature's Wild Harvest*, Blandford Press 1983.

Soothill, E., and Whitehead, P., *Wildfowl of the World*, Blandford Press 1978.

Soper, T., *The National Trust Guide to the Coast*, Webb & Bower 1984.

Yonge, C.M., *The Sea Shore* (New Naturalist Series), Collins 1949.

Index of Common Names

Page numbers in *italics* refer to illustrations.

FLORA

Index of Latin Names

SEAWEEDS

LAND SNAILS AND SLUGS

(Found near coast; no common names)

General Index